"Logic is the hygiene the mathem

"No

atics what spelling
even calligraphy) is to poetry.
Mathematical works consist
of proofs, just as poems
consist of characters."
Arnold Vladimir.

A Gentle Introduction

to

Logic, Sets

and

The Techniques of Mathematical Proofs

A Companion for High School and College Students

Brahima MBODJE, Ph.D.

AuthorHouse™
1663 Liberty Drive
Bloomington, IN 47403
www.authorhouse.com
Phone: 1-800-839-8640

First published by AuthorHouse 06/25/2011

ISBN: 978-1-4634-2967-6 (sc)
ISBN: 978-1-4634-2966-9 (ebk)

Printed in the United States of America

Library of Congress Control Number:		*2010903137*
ISBN:	*Hardcover*	*978-1-4500-5898-8*
	Softcover	*978-1-4500-5897-1*
	Ebook	*978-1-4500-5899-5*

Each problem that I solved became a rule which
served afterwards to solve other problems."
Réné Descartes

À ma fille Emma-Darbo, qui continue d'aimer son papa sans condition, dans l'espoir qu'un jour, très bientôt, elle finira par aussi aimer les mathématiques avec autant de ferveur.

"Facts do not speak. The man of science must work with method. Science is built up of facts, as a house is of stones; but an accumulation of facts is no more a science than a heap of stones is a house"
Jules Henri Poincaré (1854-1912)

TABLE OF CONTENTS

"Experience shows that it is unwise to expect much mathematical background in the case of the average student entering college. Many dread mathematics. They should be assured that mathematics is not so difficult, and that it will prove interesting if carefully studied."
American Mathematical monthly,
Vol. 40, (1933) p.391

PREFACE

Everyone is aware of the importance of proofs in mathematics. Yet, if you are a teacher of mathematics, you must have witnessed in recent years an almost total elimination of proofs from much of high-school and undergraduate mathematics curricula. The reason frequently put forward for this is that rigorous proof writing is generally a pretty hard task, and should therefore be postponed until the student has acquired sufficient mathematical maturity.

Thus, instead of teaching rigorous mathematics, many instructors resort mostly to plausibility arguments. And the bulk of students welcome this, either because they do not know what an acceptable proof in mathematics should be, or because they have always dreaded doing proofs and are therefore fine with "cutting corners." But, as a teacher, I have seen with my naked eyes what the consequences of such a neglect of rigor are; they are a rickety mathematical foundation leaving far too many students unprepared not just for graduate school mathematical works but also for upper undergraduate studies in mathematics. I am not saying that plausibility or heuristic arguments in mathematics aren't valuable, but a student of mathematics should learn to distinguish such arguments from a real mathematical proof. Thus, where-ever such heuristic arguments are used he must not only be aware of it, but he should also be encouraged to seek a more satisfactory and sounder proof.

The fact of the matter is that, contrary to wide spread belief, the skills and techniques required for successfully reading and doing rigourous proofs are not that difficult to acquire, provided one is determined to devote to their study a reasonable amount of time. Indeed, evidence shows that much of the fear students experience in learning to write proofs can be dissipated, if these students are systematically and gradually introduced to the art and strategies of proof writing.

My experience teaching the subject of mathematics for more than a decade at various levels [in colleges and secondary schools] has convinced me of a simple fact which is that even some of the most subtle techniques of proof [used by professional mathematicians] can be acquired by students of average intelligence. My goal in this book is to help this category of students to do just that. Thus, the prerequisites for reading this book are pretty modest: (1) a sound understanding of high-school algebra, (2) a reasonable exposure to *mathspeak*, (3) and above all, a strong willingness to learn mathematics.

Many a time, the obstacle before the student isn't his lack of intelligence, but the confusion he experiences with the terminology used by the mathematician. I believe that if those terms where made clear, then many more students will feel at ease with the subject. By a reasonable exposure to *mathspeak* I mean, for example, that it is imperative that the reader be able to clearly distinguish among the notions of a **definition**, an **axiom**, and **a theorem**.

- A *definition* is an accurate description of a concept; **it requires no proof whatsoever**. A definition should be given in words that are simpler than the term being defined. A valid definition in mathematics should state all that is necessary about the term and exclude all that is unnecessary. For example, a definition of a cat must include all kinds of cats, but exclude all other objects, such as dogs, houses, sheep, goats, etc., that aren't cats. Finally a correct definition is always reversible, even when its reverse is not explicitly included in the statement of the definition. Thus, the definition "A sphere is the collection of all points at equal distances to a given fixed location" should be construed to also mean "The collection of all points at equal distances to a given fixed location is a sphere." Similarly, when used as a definition, the conditional sentence

  " It is a triangle, **if** it is a three-sided plane figure" (1)

  should be understood more correctly as

"It is a triangle, **if and only if** it is a three-sided plane figure"

which includes not just (1) but also the converse of (1) given below:

"It is a three-sided plane figure, **if** it is a triangle."

It seems, therefore, that the reason one would prefer to use "if" in the place of the "if and only if" in a definition is that it is cumbersome [if not pedantic] to keep repeating the phrase "if and only if."

- An *axiom* or *postulate* is any statement which in mathematics is accepted without proof. In many cases, axioms are viewed as self-evident thereby justifying the fact that they require no proof. In all other cases, the use of axioms is a deliberate attempt to assign to the mathematical objects being studied a minimal number of properties without which the study would be impossible.
- A *theorem* is **a statement that has been proved; in other words, theorems require proofs**. To establish the proof of a theorem, one is allowed to use only definitions, axioms [that have been previously introduced], and those theorems which have already been proved; but nothing else. In general, every theorem is important and should therefore be thoroughly understood and if possible memorized, since it is likely to be used to prove another theorem. However, some less important theorems will appear under the title of *lemmas*, *corollaries*, or even as *remarks*. A lemma is a theorem whose purpose is mainly to be used in the proof of another theorem. Thus, the introduction of a lemma prior to the theorem in whose proof the lemma is intended to be used can help structure and shorten the proof of that theorem. A corollary is a theorem that is derived from another theorem with little or no additional effort. As for a remark, it is merely a [true] comment derived from a definition, axiom or theorem.

In other words, our goal in this book has been to offer the student a smooth and painless transition from high-school algebra to university mathematics. Our emphasis is not so much on the abundance of [mathematical] concepts introduced here [many of which should be familiar to many budding mathematical geniuses or high-school seniors,] but on the approach employed. Mathematics being for the most part based on deductive reasoning, at least when it comes to writing proofs, we strongly believe that the student ought not be kept in the dark about this fact. Consequently, a distinctive feature of our book is not to rush through a massive amount of material, but to spend the amount of time necessary to help readers acquire the way of thinking of the mathematician.

Hence, we won't be encyclopedic. Besides, the limited space available to us, does not permit that. Instead, we believe our choice of topics is not only right for our purpose, but also constitutes all that a serious high-school senior should master in order to be certain to succeed as a college or university mathematics major.

In addition to what I have said earlier, another important feature of this book is that it is truly self-contained; the pace of exposition is very gradual and gentle. Abundant worked examples are included to give the reader full illustrations of the concepts discussed.

This book is therefore suited for senior high-school students and college undergraduates who wish to not just do mathematics, but to also genuinely understand its language, concepts and techniques. In fact, this book should be an excellent companion for the student of mathematics, since the methods of reasoning it teaches are the ones utilized in all areas of modern [abstract] mathematics.

The present book could also serve as a textbook for a course in elementary set theory and related topics, or as one for a course on the foundation of mathematics at the undergraduate level.

The book could also be of great use to secondary school teachers of mathematics who are teaching themselves in the foundations of abstract mathematics.

To benefit fully, we advise that the reader works through the book from beginning to end, attempting the worked examples [before looking at their solutions], and redoing the proofs of theorems, lemmas and corollaries [until he finally knows how to do them on his own]. When finished, he will see how much he has demystified the notion that doing proof is difficult. This advice is even more useful for those planning to study the book on their own. We hope that the material presented in this book will serve to whet your appetite for further study of this vast and extremely import field of mathematics. Detailed solutions to all problems in the book will be posted on my webpage as soon as possible. **[However, we strongly encourage the learner to attempt all star-marked * problems at the end of every chapter**.]

I have made every reasonable effort to see to it that the book is error-free. However, *no man is perfect*, so too no man's work is perfect; therefore I should be most grateful to any reader who would bring any imperfection or suggestion to my attention.

Now, perhaps, I must reveal my deepest hope. I confess that if I have written this book, it is also for my daughter with two goals in mind; the first being to introduce her one day hopefully, with the least possible pain, to the beauties and unity of modern mathematics, and in so doing to also supplement her *mathematical French* with some *mathematical English*. So too, I would like to believe that whatever a loving father wishes for her daughter must be good for other young folks.

Finally, there is a wisdom that goes like this: *No man is an island*. So I have benefitted from the moral support of people too many for me to list here; but I would like to thank especially my mom, *Kaddy Affoue*, to whom I owe almost everything and my brother *Ousainou*. Their affection and moral support for me during the writing period of this book was crucial.

Brahima Mbodje,
Charlotte, NC
December, 2009

"A mathematician's search is fivefold: truth, usefulness, beauty, conciseness, and rigor"
The author

CHAPTER 1

PROPOSITIONAL LOGIC

1.1. PROPOSITIONS

DEFINITION

In their daily conversations people often use sentences. Some of these sentences are true; others false. Others still are neither true nor false. A Sentence that is true is said to have the truth value: **T**. And similarly, a sentence that is false is said to have the truth value: **F**.

Propositional logic is the study of sentences that are either true or false, but not both true and false. We shall call such sentences propositions.

Definition 1: A **proposition** is any declarative sentence which has one and only one of the truth values: T or F.

Example 2: Which of the following sentences are propositions? Indicate the truth value of each proposition.

2.1. Please, drive safely, or do not drive at all.
2.2. From now on.
2.3. New York City is the capital city of France.
2.4. $5+6=11$.
2.5. $x^2-25=0$.

Solution
Sentence (2.1) is not a proposition, simply because it is not a declarative; it is a request [or an interrogative sentence] instead. As for (2.2), it is not really a sentence to begin with; it is only a sentence fragment, and therefore does not qualify as a proposition. Both (2.3) and (2.4) are propositions; and clearly, (2.3) is false, whereas (2.4) is true. Now, even though (2.5) is a declarative sentence, it still does not qualify as a proposition; the reason for this is that (2.5) is neither true or false; indeed, it has no truth value until and unless one has assigned the variable x a specific number value. ♥

Remark 3: It is well to emphasize here that for a sentence to be a proposition it is not necessary that one personally has knowledge of which one of the two truth values the sentence has. What matters is that we are sure the sentence possesses exactly one of these truth values. For example, despite the fact that no one, up to the present time, can tell the actual truth value of the sentence

"There is life elsewhere beyond our solar system"

we have no doubt that this sentence is a proposition. For indeed, it is obvious that this sentence has one and only one of the two truth values. In other words, either there is life or there is no life beyond the solar system. [Our sentence is either true or false, even if we have no knowledge of its actual truth value at this point in time.] ■

As you are aware, letters such as x, y, and z are frequently used to represent numbers in algebra. So too, in logic we shall use letter symbols such as $\boldsymbol{p}$, $\boldsymbol{q}$, $\boldsymbol{r}$, $\boldsymbol{P}$, $\boldsymbol{Q}$ and $\boldsymbol{R}$ to symbolize propositions. We also know that a constant in algebra is a symbol that takes on a fixed number value. Similarly, the term **propositional constant** will be used in *logic* to refer to a proposition $\boldsymbol{P}$ whose truth value is known in advance. On the other hand, we shall use the phrase **propositional variable** to designate any proposition whose actual truth value is unknown.

SIMPLE AND COMPOUND PROPOSITIONS

In algebra, we are familiar with the *four basic operations* of addition, subtraction, multiplication and division. These operations allow us to combine numbers, say x and y, into new numbers $x+y$, $x-y$, $x \times y$, and x/y. Similarly, in logic, we may use any of the following connectives

…, and …,

…, or …,

if …, then …,

or

…, if and only if … .

to combine any two propositions into a more complex one. In order to do this, all we need is to replace both ellipsis "…" in our chosen connective with propositions.

For example, with the connective "…, **and**…" we can form the new proposition:

New York City is the capital city of France, **and** $5+6=11$.

Had we chosen another connective, say "**if**…, **then** …" instead, we would have obtained the proposition:

If New York City is the capital city of France, **then** $5+6=11$.

These are just two examples of **compound propositions.**

Definition 4: A **compound proposition** is any proposition formed by joining together two propositions with the aid of a connective. Any proposition that is not a compound proposition is called a **simple proposition**.

Remark 5: One way to tell whether a given proposition is compound or simple is to search for connectives in that proposition. If we find in it any of the above connectives [or any phrase synonymous to a connective], then the proposition is compound. ■

Example 6: Which of the following propositions are compound propositions?

6.1. Pure water is transparent and odorless.
6.2. A car is a vehicle.
6.3. Thomas Jefferson was once a president.
6.4. $5+6 \leq 2$
6. 5. Bill Clinton is an American citizen, if and only if Bill Clinton was born in the US.

Solution
Proposition (6.1) is compound, since it clearly means

Pure water is transparent, **and** pure water is odorless

which is obviously the combination of two simpler propositions by the aid of the connective "and". Propositions (6.2) and (6.3) contain no connectives [nor any terms synonymous to a connective;] they are therefore simple propositions. Proposition (6.4) is a compound proposition, simply because "$5+6 \leq 2$" stands for "$5+6<2$ or $5+6=2$," a form that involves the connective "or". Proposition (6.5) too is a compound proposition, since it is the combination of two propositions with the connective "if and only if":

Bill Clinton is an American citizen, **if and only if** Bill Clinton was born in the US. ♥

1.2. TYPES OF COMPOUND PROPOSITIONS

We saw in the preceding section that propositions may be classified into two categories: simple propositions, and compound propositions. In what follows, we will see that the category of compound propositions can itself be divided into four subcategories: **conjunctions**, **disjunctions**, **conditionals**, and **biconditionals**.

CONJUNCTIONS

Definition 7: A compound proposition is called a **conjunction**, if it can be formed by joining together two propositions with the aid of the connective "…, **and** …."

Now that we have defined a conjunction, the good question to ask ourselves is: **how does one determine the truth value of a conjunction from the truth values of its constituent propositions?** The answer to this question is given by the following convention.

Axiom 8: Let p and q be any two propositions. Their conjunction, symbolized by $p \wedge q$ and read p **and** q, is considered to be true, when and only when its constituent propositions, p and q, are both true.

Since there are two possible truth values for any propositional variable, there must be four possible truth value combinations for any two propositional variables p and q taken together. We list these combinations below in Table 1.1:

p	q
T	T
T	F
F	T
F	F

Table 1.1

For each of these combinations we can easily determine the corresponding truth value of the conjunction $p \wedge q$ by using the foregoing axiom. The result is Table 1.2 called the truth table for $p \wedge q$. [Note that the combination for which the conjunction is true is shown in bold-type.]

p	q	$p \wedge q$
T	**T**	**T**
T	F	F
F	T	F
F	F	F

Table 1.2

Example 9: Determine the truth values of the following conjunctions

9.1. Pure water is transparent and odorless.
9.2. A triangle is a four-sided plane figure with 3 vertices.
9.3. $5+6>12$, and $5<5$.
9.4. Hilary Clinton was elected as a US senator, though she was a woman.

Solution
Conjunction (9.1) is true, since both of its components, "*Pure water is transparent*" and "*Pure water is odorless*", are true. Conjunction (9.2) is false; indeed, if rewritten as "*A triangle is a four-sided plane figure, and a triangle has 3 vertices*", then it is clear that not both of its constituent propositions are true. Conjunction (9.3) is also a false proposition, since not both of its constituent propositions are true; in fact, none of them is true. As for conjunction (9.4), it may be rewritten as "*Hilary Clinton was elected as a US senator, and Hilary Clinton was a woman*"; thus it is a true conjunction, since both of its constituent propositions are true. ♥

DISJUNCTION

Definition 10: A compound proposition is called a **disjunction**, if it can be formed by joining together two propositions with the aid of the connective "…, **or**…."

Next is the convention for deciding the truth values of disjunctions.

Axiom 11: Let p and q be any two propositions. Their disjunction, symbolized by $p \vee q$ and read p **or** q, is considered to be true, when and only when at least one of the two constituent propositions is true.

Using this axiom we readily obtain the following truth table for $p \vee q$.

p	q	$p \vee q$
T	T	T
T	F	T
F	T	T
F	F	F

Table 1.3

Example 12: Determine the truth values of the following disjunctions

12.1. Pure water is transparent or odorless.

12.2. A triangle is a four-sided plane figure or a figure that has 3 vertices.

12.3. $5 + 6 > 12$, or $5 < 5$.

12.4. Hilary Clinton was elected as a US senator, or she was a woman.

Solution

Disjunction (12.1) is true since at least one of its constituent propositions is true; in fact both of them are true. Disjunction (12.2) is also true, because at least one of its constituent propositions, namely "*A triangle has 3 vertices*", is true. Disjunction (12.3) is a false proposition, because none of its constituent proposition is true. Disjunction (12.4) is true, since at least one of its constituent propositions is true; in fact both of its constituent propositions are true. ♥

CONDITIONALS

Definition 13: A compound proposition is said to be a **conditional**, if it is formed by joining together two propositions with the aid of the connective "**If…, then….**"

To decide the truth value of a conditional, mathematicians and logicians use the following convention.

Axiom 14: Let p and q be any two propositions. The conditional of p and q, symbolized by $p \to q$ and read **if p then q**, is true for all truth value combinations of p and q, except where p is true and q is false.

In a conditional $p \to q$, the proposition p that precedes the arrow is called the **hypothesis**, whereas q, the proposition that follows the arrow, is referred to as the **conclusion**. With these terminologies in mind, we may rephrase Axiom 14 as follows:

Axiom 15: Let p and q be any two propositions. Then the conditional $p \to q$ is true for all truth value combinations of p and q, except where the hypothesis is true and the conclusion is false.

Hence, the truth table for $p \to q$ is as follows:

p	q	$p \to q$
T	**T**	**T**
T	F	F
F	**T**	**T**
F	**F**	**T**

Table 1.4

Remarks 16: We deduce from the last two rows of Table 1.4 that conditionals with false hypothesis are automatically true, no matter what the truth values of their conclusions may be. Similarly, we see from rows 2 and 4 that any conditional whose conclusion is true is also automatically true. ■

Example 17: Determine the truth values of the following conditionals

17.1. If Socrates is a man, then Socrates is mortal
17.2. If Paris is as little as a fly, then Paris can fit in a bottle.
17.3. If $5+6>12$, then $5=5$.
17.4. If $5=5$ then $5+6>12$.

Solution
Observe that of all four conditionals, (17.4) is the only one that combines a true hypothesis "$5=5$" with a false conclusion "$5+6>12$". Thus, by virtue of Axiom 15, (17.4) is the only false conditional here. All the remaining conditionals are therefore true. ♥

BICONDITIONALS

Definition 18: A compound proposition is called a **biconditional**, if it is formed by joining together two propositions with the aid of the connective "..., **if and only if**...."

To get the truth value of a biconditional we shall apply the following convention.

Axiom 19: Let p and q be any two propositions. Then the biconditional of p and q, symbolized by $p \leftrightarrow q$ and read p **if and only if** q, is considered to be true, when and only when both p and q have exactly the same truth values.

Thus, the truth table for $p \leftrightarrow q$ is:

p	q	$p \leftrightarrow q$
T	**T**	**T**
T	F	F
F	T	F
F	**F**	**T**

Table 1.5

Example 20: Determine the truth values of the following biconditionals

20.1. Socrates is a Greek, if and only if Socrates was born in Greece
20.2. Paris is as small as a fly, if and only if Paris can fit in a bottle.
20.3. $5+6>12$, if and only if $5=5$.
20.4. $3>1000$, if and only if $34>12$.

Solution
The biconditional (20.1) is true, since both of its constituent propositions, "*Socrates is a Greek*" and "*Socrates was born in Greece*", are true. The biconditional (20.2) is also true, because the components propositions, "*Paris is as small as a fly*" and "*Paris can fit in a bottle*", are both false. The biconditional (20.3) is false, since its constituent propositions, "$5+6>12$" and "$5=5$", have opposite truth values. Similarly, the biconditional (20.4) is false, since its component propositions, "$3>1000$" and "$34>12$", do not share the same truth value. ♥

1.3. DENIAL OF A PROPOSITION

Sometimes, we feel the need to deny a proposition, all the more so if we believe it to be false. For instance, upon hearing the following proposition "*Parrots are insects*", which is obviously a false proposition, we may feel an urge to voice the denial: "*It is not the case that parrots are insects*".

Definition 21: Let p be any proposition. The **denial** of p, symbolized by $\sim p$ and read **not** p, is that new proposition obtained by prefixing p with the phrase **"It is not the case that..."** or with any other synonymous phrase.

Example 22: Form the denials of these two propositions

22.1. London is the capital city of Great Britain.

22.2. $1+1>10$

Solution

The denial of proposition (22.1) is

"It is not the case that London is the capital city of Great Britain"

which is obviously the same as

"London **is not** the capital city of Great Britain."

Similarly, the denial of (22.2) is

"It is not the case that $1+1>10$"

or simply

" $1+1 \leq 10$." ♥

Remark 23: It is not difficult to see that the denial of a false proposition is a true proposition; and that the denial of a true proposition results in a false proposition. In other words, a proposition and its denial always have opposite truth values. ■

Thus, the truth table for $\sim p$ is simply:

p	$\sim p$
T	F
F	T

Table 1.6

Example 24: Draw the truth table of the double denial $\sim(\sim p)$. What can you notice?

Solution

p	$\sim p$	$\sim(\sim p)$
T	F	**T**
F	T	**F**

Table 1.7

We notice that p and its double denial $\sim(\sim p)$ share the same truth value. We shall convey this fact by saying that the two propositions p and $\sim(\sim p)$ are **logically equivalent**. ♥

1.4. PROPOSITIONAL EXPRESSIONS

In algebra, we can combine number variables x and y with binary operations to form algebraic expressions such as $x^2 + 2(x + y) + 7y^6$. So too, in logic we may combine propositional variables p, q, and r with logical connectives to obtain, say, the following **propositional expressions**:

$$[(\sim p) \vee q] \rightarrow (p \leftrightarrow q). \quad (1)$$

Definition 25: A **propositional expression** in logic is any **meaningful** combination of propositional variables with the aid of logical connectives.

Example 26: Which of the following combinations are propositional expressions?

26.1. $[\vee[(\sim p) \vee q]] \rightarrow p$

26.2. $[p \vee q] \rightarrow [p \wedge (\sim q)]$

26.3. $\sim[\sim(\sim q)]$

26.4. $[\leftrightarrow q\,] \vee p$

Solution
The combination (26.1) is not a propositional expression, since $\vee\,[\,(\,\sim p\,) \vee q\,]$ makes no sense at all. On the other hand, both (26.2) and (26.3) are propositional expressions, since both of them make perfect sense. Finally, the combination (26.4) is not a propositional expression; for indeed, $\leftrightarrow q$ does not make any sense at all. ♥

Remark 27: In case you have difficulties trying to figure out whether or not a given combination constitutes a propositional expression, just establish its truth table. If you find out that it is impossible to do so, then your combination is not a propositional expression. ■

Now, note that in (1) we used multiple pairs of brackets. This was done deliberately to indicate the order in which the connectives should be applied. Fortunately though, such profuse [and therefore cumbersome] use of brackets may be avoided, provided we adopt some convention:

Convention 28 [Order of connectives]: In propositional expressions, denials and connectives are to be applied according to the following order:

1st : Connectives within brackets; innermost brackets are handled first.
2nd : $\sim$
3rd : $\wedge$ or $\vee$ in the order in which they appear from left to right.
4th : $\rightarrow$
5th : $\leftrightarrow$

Thus, by adopting the foregoing convention, we can simplify the propositional expression (1) as:

$$\sim p \vee q \rightarrow (p \leftrightarrow q).$$

Exercise 29: Suppose p is a true proposition, and q a false proposition. Find the truth value of the propositional expression $p \vee q \rightarrow p \wedge \sim q$.

Solution

The truth value of $p \vee q \rightarrow p \wedge \sim q$ is provided in the last column of the following table:

p	q	$\sim q$	$p \vee q$	$p \wedge \sim q$	$p \vee q \rightarrow p \wedge \sim q$
T	F	T	T	T	T

Table 1.8

With p true and q is false, we have found, by applying the preceding convention, that $p \vee q \rightarrow p \wedge \sim q$ is true. ♥

Example 30: Which of the following are propositional expressions?

30.1. $\vee \sim p \vee q \rightarrow p$
30.2. $p \vee q \rightarrow p \wedge \sim q$
30.3. $\sim\sim\sim q$
30.4. $p \wedge \sim \vee \leftrightarrow q \vee p$

Solution

The combination (30.1) is not a propositional expression, since $\vee \sim p$ makes no sense at all. On the other hand, combination (30.2) is a propositional expression; since it makes perfect sense. The combination (30.3) too is a propositional expression; indeed, it signifies $\sim \{\sim (\sim q)\}$ which makes perfect sense. As for combination (30.4), it is not a propositional expression, since $p \wedge \sim \vee \leftrightarrow q$ makes no sense at all. ♥

We are now in a position to introduce the useful concept of the **main connective** in a propositional expression.

Definition 31: By the **main connective** in a given propositional expression, we mean that connective which is to be applied last.

The reason this is an important concept is that it will allow us to classify propositional expressions into different types: **denial**, **conjunction**, **disjunction**, **conditional** or **biconditional**.

Example 32: Indicate whether the following propositional expressions are denial, conjunctions, disjunctions, conditionals or biconditionals.

32.1. $\sim(p \rightarrow p)$
32.2. $q \rightarrow (t \leftrightarrow \sim q)$
32.3. $\sim q \vee p$
32.4. $p \leftrightarrow q \vee p$
32.5. $(p \leftrightarrow p) \wedge r$

Solution
Applying Convention 28 to these expressions, we obtain:

Propositional Expression	Last connective	Type of Proposition
$\sim(p \rightarrow p)$	$\sim$	Denial
$q \rightarrow (t \leftrightarrow \sim q)$	$\rightarrow$	Conditional
$\sim q \vee p$	$\vee$	Disjunction
$p \leftrightarrow q \vee p$	$\leftrightarrow$	Biconditional
$(p \leftrightarrow p) \wedge r$	$\wedge$	Conjunction

Table 1.9

which completes the solution . ♥

1.5. CONSTRUCTION OF TRUTH TABLES

In the preceding sections we constructed truth tables for a number of propositional expressions. These tables were easy to arrive at, because the expressions considered involved no more than two propositional variables. However, for more complex propositional expressions, such as this one:

$$[p \wedge \sim s \leftrightarrow r] \rightarrow s, \tag{2}$$

which involve several connectives and three or more propositional variables, the task of constructing a correct truth table may be a frustrating one for beginners. The frustration can however be alleviated, if one proceeds in a systematic step by step manner. We will demonstrate this with propositional expression (2).

Step 1. Start by finding out how many distinct propositional variables are involved in our expression. In the case here, there are 3 different propositional variables: $\boldsymbol{p}$, $\boldsymbol{r}$, and s. Since each variable has 2 possible truth values, the total number of truth value combinations for the 3 propositional variables [considered together] is: $2 \times 2 \times 2 = 2^3 = 8$. These are shown in the table below.

p	r	s
T	T	T
T	T	F
T	F	T
T	F	F
F	T	T
F	T	F
F	F	T
F	F	F

Table 1.10

Step 2. Now add a new column to the right of Table 1.10. This column is for the connective in (2) that should be applied **first** according to the order of operations. We should have:

p	r	s	$\sim s$
T	T	T	F
T	T	F	T
T	F	T	F
T	F	F	T
F	T	T	F
F	T	F	T
F	F	T	F
F	F	F	T

Table 1.11

Step 3. Add another column to the right of Table 1.11. This column is for the connective in (2) that should be applied **second** according to the order of operations. We have:

p	r	s	$\sim s$	$p \wedge \sim s$
T	T	T	F	F
T	T	F	T	T
T	F	T	F	F
T	F	F	T	T
F	T	T	F	F
F	T	F	T	F
F	F	T	F	F
F	F	F	T	F

Table 1.12

Step 4. Add another column to the right of Table 1.12. This column is for the connective in (2) that should be applied **third** according to the order of operations. We should arrive at:

p	r	s	$\sim s$	$p \wedge \sim s$	$p \wedge \sim s \leftrightarrow r$
T	T	T	F	F	F
T	T	F	T	T	T
T	F	T	F	F	T
T	F	F	T	T	F
F	T	T	F	F	F
F	T	F	T	F	F
F	F	T	F	F	T
F	F	F	T	F	T

Table 1.13

Step 5. Add yet another column to the right of Table 1.13. This column is for the main connective, the connective in (2) that should be applied **last** according to the order of operations. We have:

p	r	s	$\sim s$	$p \wedge \sim s$	$p \wedge \sim s \leftrightarrow r$	$[p \wedge \sim s \leftrightarrow r] \rightarrow s$
T	T	T	F	F	F	T
T	T	F	T	T	T	F
T	F	T	F	F	T	T
T	F	F	T	T	F	T
F	T	T	F	F	F	T
F	T	F	T	F	F	T
F	F	T	F	F	T	T
F	F	F	T	F	T	F

Table 1.14

Example 33: Draw the truth table of the propositional expression: $p \wedge q \rightarrow p$. What do you notice?

Solution:

p	q	$p \wedge q$	$p \wedge q \rightarrow p$
T	T	T	**T**
T	F	F	**T**
F	T	F	**T**
F	F	F	**T**

Table 1.15

We notice that the propositional expression $p \wedge q \rightarrow p$ is always true, and this, no matter the truth or falsity of its component propositions p and q. ♥

PROBLEMS

1. Use Convention 28, about the order of connectives, to add parentheses to each of the following propositional expressions. Then indicate whether each propositional expression is a denial, conjunction, disjunction, conditional, or biconditional.

   1.1. $p \rightarrow p \wedge q$
   1.2. $\sim p \wedge q \rightarrow r \vee q$
   1.3. $\sim r \vee \sim q \leftrightarrow p \rightarrow r$
   1.4. $\sim r \vee \sim q \rightarrow p$
   1.5. $p \wedge \sim p$
   1.6. $\sim r \rightarrow \sim q \vee p$.

2. Construct a truth table for each of the following propositional expressions.
   2.1. $p \wedge \sim p$
   2.2. $\sim p \vee q$
   2.3. $\sim (p \vee \sim p)$
   2.4. $\sim p \wedge (r \vee \sim q)$
   2.5. $(\sim r \vee \sim q) \rightarrow p$
   2.6. $(\sim r \vee \sim q) \leftrightarrow (p \rightarrow r)$.

3. * Construct one truth table for both of these propositional expressions, and observe that if the first one is true then so is the second.

$$[a \rightarrow (p \vee q)] \wedge [p \rightarrow b] \wedge [q \rightarrow b] \quad \text{and} \quad a \rightarrow b.$$

"Logic may be defined as the art of drawing inferences; in an important sense everybody draws inferences; even animals do so. But most people's inferences are rash and hasty; subsequent experience shows them to be wrong. Logic aims at avoiding such unreliable kinds of inference; it is analogous to the rules of evidence in law"
Bertrand Russell .

CHAPTER 2

TAUTOLOGIES AND CONTRADICTIONS

2.1. TAUTOLOGIES AND CONTRADICTIONS

TAUTOLOGY

At the end of the preceding chapter, in Example 33, we saw that the propositional expression

$$p \wedge q \rightarrow p$$

is true regardless of the truth or falsity of its constituent propositional variables p and q. From now on, we will refer to this fact by saying that $p \wedge q \rightarrow p$ is a **tautology**.

Definition 1: In general, any propositional expression, such as the one above, which is true for every truth combination of its constituent variables, is called a **tautology**. We shall denote tautologies by ***TAUT***.

Remark 2: Thus, it is clear that a tautology is a propositional constant with the truth value T. ■

There are other tautologies. And, unlike the one we just mentioned above, not all tautologies are conditionals

Example 3: Show that the [disjunctional] propositional expression $p \vee \sim p$ is also a tautology.

Solution:
To show that $p \vee \sim p$ is a tautology we construct its truth table.

p	$\sim p$	$p \vee \sim p$
T	F	**T**
F	T	**T**

Since $p \vee \sim p$ is always true regardless of the truth or falsity of the propositional variable p, we conclude that $p \vee \sim p$ is indeed a tautology. ♥

Example 4: Show that the conditional [propositional] expression $p \rightarrow TAUT$ is a tautology.

Solution:
Again, to show that $p \rightarrow TAUT$ is a tautology we construct its truth table.

p	$TAUT$	$p \rightarrow TAUT$
T	T	**T**
F	T	**T**

Since $p \rightarrow TAUT$ is true regardless of the truth or falsity of its propositional variable p, we conclude that $p \rightarrow TAUT$ too is a tautology. ♥

CONTRADICTION

Now, consider the expression $\sim(p \vee \sim p)$. From Example 3 above, we saw that $p \vee \sim p$ is true regardless of the truth or falsity of its propositional variable p. Therefore, $\sim(p \vee \sim p)$, the denial of $p \vee \sim p$, must always be false regardless of the truth or falsity of p.

Definition 5: In general, any expression, such as $\sim(p \vee \sim p)$, which is false for every truth combination of its propositional variables is called a **contradiction**. We shall denote contradictions by ***CONT***.

Remark 6: It is clear from this definition that a contradiction is merely a propositional constant with the truth value F. Also, it is easy to see that the denial of a tautology is a contradiction. ■

Example 7: Show that the expression $\sim(CONT \rightarrow p)$ is a contradiction.

Solution:
To show that the propositional expression $\sim(CONT \rightarrow p)$ is a contradiction we may first construct its truth table.

$CONT$	p	$CONT \rightarrow p$	$\sim(CONT \rightarrow p)$
F	T	T	**F**
F	F	T	**F**

Since $\sim(CONT \rightarrow p)$ is false regardless of the truth or falsity of the propositional variable p, we conclude that $\sim(CONT \rightarrow p)$ is indeed a contradiction. ♥

2.2. IMPLICATIONS AND EQUIVALENCES

We will now introduce two very important concepts: **implications** and **equivalences**.

IMPLICATION

Definition 8: Let R and **S** be two propositional expressions. We shall say that R **implies** S, in which case we write $R \Rightarrow S$, if the conditional $R \rightarrow S$ is a tautology.

Next, we make two important remarks from the foregoing definition.

Remark 9: Let's assume that

$$R \Rightarrow S.$$

Then from Definition 8,

$$R \to S \text{ is a tautology.}$$

Thus

If R is true, then S too is true. ■

Remark 10: Conversely[1], the assumption that

If R is true, then S is also true

means, by Remarks 16 of Chapter 1, that the conditional

$$R \to S$$

can never be false. Hence,

$$R \to S \text{ is a tautology,}$$

which, in turn signifies that

$$R \Rightarrow S. \blacksquare$$

From these two remarks, we therefore immediately have

Theorem 11: " $R \Rightarrow S$ " is the same as "Whenever R is true, then so too is S".

Example 12: Show that $[\,(p \to q) \wedge (q \to r)\,] \Rightarrow (p \to r)$.

Solution:
To show the above implication, we will first construct a truth table for both of the propositional expressions $(p \to q) \wedge (q \to r)$ and $p \to r$.

[1]. The term **conversely** is frequently used in mathematics to sound an advance warning to the reader that an argument just made is about to be reversed, and that the reverse argument also holds true.

p	q	r	$p \to q$	$q \to r$	$(p \to q) \wedge (q \to r)$	$p \to r$
F	F	F	T	T	**T**	**T**
F	F	T	T	T	**T**	**T**
F	T	F	T	F	F	**T**
F	T	T	T	T	**T**	**T**
T	F	F	F	T	F	F
T	F	T	F	T	F	**T**
T	T	F	T	F	F	F
T	T	T	T	T	**T**	**T**

Now, notice from the above truth table that whenever the expression $(p \to q) \wedge (q \to r)$ is true, then so too is the propositional expression $p \to r$. Thus, by Theorem 11, we conclude that

$$[\,(p \to q) \wedge (q \to r)\,] \Rightarrow (p \to r). \quad ♥$$

Example 13: Show that

$$[a \to (p \vee q)] \wedge [p \to b] \wedge [q \to b] \Rightarrow [a \to b]\,.$$

Solution:

First, we must construct one truth table for both

$$[a \to (p \vee q)] \wedge [p \to b] \wedge [q \to b]$$

and

$$a \to b.$$

You may refer to Problem 3 of Chapter 1. There, you should find that whenever $[a \to (p \vee q)] \wedge [p \to b] \wedge [q \to b]$ is true, so too is $a \to b$. Hence, invoking Theorem 11, we have:

$$[a \to (p \vee q)] \wedge [p \to b] \wedge [q \to b] \Rightarrow a \to b. \quad ♥$$

Therefore, recapping the results of the last two examples, we arrive at these two very important theorems:

Theorem 14 [Law of transitivity]

Let p, q and r be any three propositions. Then:

$$(p \to q) \wedge (q \to r) \Rightarrow (p \to r).$$

and

Theorem 15 [Law of exhaustion of all cases]:

Suppose a, p, q, and r are any four propositions. Then

$$[a \to (p \vee q)] \wedge [p \to b] \wedge [q \to b] \Rightarrow (a \to b).$$

EQUIVALENCES

Definition 16: Let R and S be propositional expressions. We shall say that **R is equivalent to S**, in which case we write $R \Leftrightarrow S$, if the biconditional $R \leftrightarrow S$ is a tautology.

Example 17: Establish each of the equivalences below.

1. $\sim(p \wedge q) \Leftrightarrow \sim p \vee \sim q$.
2. $\sim(p \vee q) \Leftrightarrow \sim p \wedge \sim q$.

Solution:

To treat the first part of this example we first construct a truth table for both of the expressions $\sim(p \wedge q)$ and $\sim p \vee \sim q$.

p	q	$p \wedge q$	$\sim(p \wedge q)$	$\sim p$	$\sim q$	$\sim p \vee \sim q$
F	F	F	**T**	T	T	**T**
F	T	F	**T**	T	F	**T**
T	F	F	**T**	F	T	**T**
T	T	T	**F**	F	F	**F**

Note that propositions $\sim(p \wedge q)$ and $\sim p \vee \sim q$ share exactly the same truth values. Hence, $\sim(p \wedge q) \leftrightarrow \sim p \vee \sim q$ is a tautology, and so we may write,

$$\sim(p \wedge q) \Leftrightarrow \sim p \vee \sim q\ .$$

Similarly, to treat the second part of this example we must first of all construct one truth table for the propositions $\sim(p \vee q)$, and $\sim p \wedge \sim q$.

p	q	$p \vee q$	$\sim(p \vee q)$	$\sim p$	$\sim q$	$\sim p \wedge \sim q$
F	F	F	**T**	T	T	**T**
F	T	T	**F**	T	F	**F**
T	F	T	**F**	F	T	**F**
T	T	T	**F**	F	F	**F**

Here too, one can see that $\sim(p \vee q)$, and $\sim p \wedge \sim q$ have exactly the same truth values. Hence, they too are equivalent:

$$\sim(p \vee q) \Leftrightarrow \sim p \wedge \sim q. \quad \heartsuit$$

We gather the results of Example 17 into the following

Theorem 18 [De Morgan Laws of Logic]:

Suppose p and q are propositions. Then

1. $\sim(p \wedge q) \Leftrightarrow \sim p \vee \sim q,$

and

2. $\sim(p \vee q) \Leftrightarrow \sim p \wedge \sim q.$

In other words, the denial of the conjunction of any two propositions is the disjunction of their denials; and similarly the denial of the disjunction of any two propositions is the conjunction of their denials.

Example 19: Show that $p \rightarrow q \Leftrightarrow \sim q \rightarrow \sim q$.

Solution:
To show $p \rightarrow q \Leftrightarrow \sim q \rightarrow \sim p$, we will start by drawing one truth table for both of the propositions $p \rightarrow q$ and $\sim q \rightarrow \sim q$.

p	q	$p \rightarrow q$	$\sim q$	$\sim p$	$\sim q \rightarrow \sim p$
F	F	**T**	T	T	**T**
F	T	**T**	F	T	**T**
T	F	**F**	T	F	**F**
T	T	**T**	F	F	**T**

We notice that $p \rightarrow q$ and $\sim q \rightarrow \sim p$ have exactly the same truth values. Therefore,

$$p \rightarrow q \Leftrightarrow \sim q \rightarrow \sim p.$$

The conditional $\sim q \rightarrow \sim p$ is often referred to as the **contrapositive** of the original conditional $p \rightarrow q$. ♥

Again, we gather the result of this preceding example for use at a later time as

Theorem 20 [Law of contraposition]: Suppose p and q are any two propositions. Then

$$p \rightarrow q \Leftrightarrow \sim q \rightarrow \sim p.$$

In other words, every conditional is equivalent to its contrapositive.

Example 21: Show that $p \Leftrightarrow \sim p \rightarrow CONT$.

Solution:
To treat this example, we will first construct a truth table for the propositions:

$$p \text{ and } \sim p \rightarrow CONT.$$

p	$\sim p$	$CONT$	$\sim p \rightarrow CONT$
F	T	F	F
T	F	F	T

Hence, we see that p and $\sim p \rightarrow CONT$ always have exactly the same truth values. Thus, they are equivalent:

$$p \Leftrightarrow \sim p \rightarrow CONT. \quad ♥$$

Again, we gather the foregoing result into another theorem for use in later chapters.

Theorem 22 [Law of contradiction]:

Suppose p is a proposition. Then

$$p \Leftrightarrow \sim p \rightarrow CONT.$$

Example 23: Show that $p \leftrightarrow q \Leftrightarrow (p \rightarrow q) \wedge (q \rightarrow p)$.

Solution:
We first draw a truth table for the two propositional expressions $p \leftrightarrow q$ and $(p \rightarrow q) \wedge (q \rightarrow p)$.

p	q	$p \leftrightarrow q$	$p \rightarrow q$	$q \rightarrow p$	$(p \rightarrow q) \wedge (q \rightarrow p)$
F	F	**T**	T	T	**T**
F	T	**F**	T	F	**F**
T	F	**F**	F	T	**F**
T	T	**T**	T	T	**T**

We see that $p \leftrightarrow q$ and $(p \rightarrow q) \wedge (q \rightarrow p)$ have exactly the same truth values. Thus,

$$p \leftrightarrow q \Leftrightarrow (p \rightarrow q) \wedge (q \rightarrow p)$$

is indeed an equivalence. In logic, the pair of conditionals $p \rightarrow q$ and $q \rightarrow p$ are said to be **converses** of one another. ♥

We gather the result of Example 23 as another theorem:

Theorem 24 [Law of reciprocity] : Let p and q be propositions. Then

$$p \leftrightarrow q \Leftrightarrow (p \rightarrow q) \wedge (q \rightarrow p).$$

In other words, every biconditional $p \leftrightarrow q$ is equivalent to the conjunction of $p \rightarrow q$ with its converse $q \rightarrow p$.

We now wish to make a couple of remarks from Definition 16.

Remark 25: Suppose

$$R \Leftrightarrow S.$$

Then from Definition 16,

$R \leftrightarrow S$ is a tautology

which, in turn means

R and S always share a same truth value.

Thus,

[When R is true, S too is true] and [When S is true, R too is true].

Symbolically, that means

$R \Rightarrow S$ and $S \Rightarrow R$. ■

Remark 26: Conversely, the assumption that

$R \Rightarrow S$ and $S \Rightarrow R$

means

[When R is true, S too is true] and [When S is true, R too is true],

which in turn signifies that

R and S always share a same truth value.

Thus,

$R \leftrightarrow S$ is a tautology.

Therefore, by Definition 16,

$$R \Leftrightarrow S. \blacksquare$$

And so we have proved the following assertion.

Theorem 27 [Law of reciprocity]: Let R and S be propositions. Then

"$R \Leftrightarrow S$" is the same as "$R \Rightarrow S$ and $S \Rightarrow R$."

PROBLEMS

1. We saw in Theorem 20 of the present chapter that a conditional $p \to q$ and its contrapositive $\sim q \to \sim p$ are equivalent. Now, show that a conditional $p \to q$ and its converse $q \to p$ are not necessarily equivalent.
2. Show that $p \Rightarrow TAUT$.
3. Show that $CONT \Rightarrow p$.
4. * Show that $p \to q \Leftrightarrow \sim p \vee q$.
5. Show that $p \Leftrightarrow \sim(\sim p) \vee p \Leftrightarrow \sim p \to p$.
6. Show that $[p \leftrightarrow q] \Leftrightarrow [\sim p \leftrightarrow \sim q]$.
7. Show that $[p \Leftrightarrow q] \Leftrightarrow [\sim p \Leftrightarrow \sim q]$.
8. Show that $p \Leftrightarrow p \vee p$.
9. Show that $p \Leftrightarrow p \wedge p$.
10. * Show that $p \wedge (r \wedge s) \Leftrightarrow (p \wedge r) \wedge (p \wedge s)$.
11. * Show that $p \wedge (r \vee s) \Leftrightarrow (p \wedge r) \vee (p \wedge s)$.
12. * Show that $p \Rightarrow (p \vee q)$.

13. Show that $p \wedge q \Rightarrow p$.

14. * Show that $p \Leftrightarrow \sim p \rightarrow p$.

15. * Show **[The Law of The Excluded Middle]**

    $\sim p \vee p \Leftrightarrow TAUT$ and $\sim p \wedge p \Leftrightarrow CONT$.

    It is believed that this law was first articulated by Aristotle. For that reason it is also known as the law of *Aristotelian logic.*

16. * Show that

    $(p \vee q) \wedge (r \vee s) \Leftrightarrow (p \wedge r) \vee (p \wedge s) \vee (q \wedge r) \vee (q \wedge s)$.

17. Show that $CONT \vee p \Leftrightarrow p$.

18. * Show that $p \wedge q \Leftrightarrow \sim(\sim p \vee \sim q)$.

" Mathematicians create by acts of insight and intuition. Logic then sanctions the conquests of intuition."
Morris Kline (1908-1992)

"It is an old maxim of mine that when you have excluded the impossible, whatever remains, however improbable, must be the truth. "
Sherlock Holmes

CHAPTER 3

THEOREMS, AND PROOF METHODS

3.1. THEOREMS

You certainly remember the definition of a proposition; we learned in Chapter 1 that a proposition is any declarative sentence that is either true or false, but not both. We also emphasized the fact that for a sentence to qualify as a proposition, it is not necessary that we personally have knowledge of its actual truth value. And this, since all that matters is that we are certain our proposition is either true or false.

However, in the present chapter, we will be concerned with those propositions whose truth values have been proved to be **T**. Such propositions are customarily called **theorems**.

Definition 1: A **theorem** in mathematics is simply **a true proposition**. Hence , it comes in one of these three forms:

(1) a tautology P,

(2) an implication $P \Rightarrow Q$,

(3) or an equivalence $P \Leftrightarrow Q$.

The focus here is to discuss some of the methods or strategies commonly used in mathematics to prove theorems. As will be seen, the justifications for these proof strategies are the laws of logic discovered in Chapter 2.

By discussing and illustrating these various proof methods, we intend to bring the reader to a point where he will be able to follow proofs written by others, and to recognize in them every single proof strategy being employed. This is all the more important, since the ability to recognize the methods used by others [at every stage in their proofs] is very instrumental in helping us learn how to write our own proofs.

Our concern in this chapter is not so much to reveal the intricate mental process by which a mathematician is able to intuit a theorem, than to learn the methods by which these theorems once conjectured may be proved. For, it seems to us that the superiority of one person over another at conjecturing useful theorems is to be found elsewhere, perhaps in neuroscience or psychology, two fields in which we are the least knowledgeable. All we are sure of is that your ability to conjecture useful theorems is bound to increase significantly with your mathematical experience and maturity.

Innate gift, creativity, cleverness, imagination and artfulness are certainly pretty important assets for anyone wishing to learn to be good at authoring proofs; but equally important is one's willingness to study proofs by other mathematicians, and to also practise writing one's own proofs. Indeed, writing proofs in mathematics is much similar to a detective work in unraveling crime cases; and one does get better at it only through frequent practice[2].

A proof in mathematics may contain one single method, or a combination of several proof strategies. In what follows, we discuss six major proof strategies.

---

[2] The difference here is that detective work in mathematics is almost cost-free; whereas in real crimes, it often requires expensive resources that only a state or big organization can afford.

3.2. PROOF METHODS

Next, we discuss these methods or strategies, and illustrate each of them with an example.

THE METHOD OF TRIVIAL PROOFS

To establish an implication

$$p \Rightarrow q$$

using this method of proof, it suffices to show that q is a tautology. Indeed, if q is a tautology, then $p \rightarrow q$ is always true no matter the truth or falsity of p. Thus $p \rightarrow q$ too would be a tautology, which is the same as asserting that the implication $p \Rightarrow q$ holds true.

Example 2: Suppose r and s are propositions. Prove that:

$$r \Rightarrow \sim s \vee s.$$

Solution

We learned in Chapter 2 that $\sim s \vee s$ is a tautology. Hence,

$$r \Rightarrow \sim s \vee s$$

is a true implication. ♥

THE METHOD OF VACUOUS PROOFS

To prove an implication

$$p \Rightarrow q$$

using this method, it suffices to establish that p is a contradiction. Indeed, if p is a contradiction, then $p \rightarrow q$ is always true no matter the truth value of q. And this would mean that $p \rightarrow q$ is a tautology, which in turn is the same as saying that $p \Rightarrow q$ holds true.

Example 3: In set theory, the **empty set**, or the set which contains no element whatsoever, is denoted by $\varnothing$. One also defines the idea of a **subset** by saying that a set A is a subset of another set B, in which case one writes $A \subset B$, if every item in A is also an item of B.

Having introduced these two notions, we will now prove that the empty set $\varnothing$ is a subset of any given set S.

Solution
Using the definitions of the empty set and a subset, it is easy to see that the problem before us is to prove the implication:

$$x \in \varnothing \Rightarrow x \in S \tag{1}$$

But, note that $x \in \varnothing$ is a contradiction. Indeed, since $\varnothing$ is empty, there is no way it can contain any item x. Hence, using Remarks 16 of Chapter 1, we conclude that the conditional

$$x \in \varnothing \rightarrow x \in S.$$

is a tautology. Thus, (1) is true. ♥

THE METHOD OF DIRECT PROOFS

To prove

$$p \Rightarrow q$$

using this method, it suffices to exhibit a chain of implications arranged from (the hypothesis) p to (the conclusion) q:

$$q \Rightarrow r_1 \Rightarrow r_2 \Rightarrow r_3 \Rightarrow \ldots \Rightarrow r_{n-1} \Rightarrow p \tag{2}$$

where $r_1, r_2, r_3, \ldots,$ and r_{n-1} are a finite number of propositions. Indeed, suppose we are able to establish such a chain of implications. Then, we have the following chain of conditionals, each of which is a tautology:

$$q \rightarrow r_1 \rightarrow r_2 \rightarrow r_3 \rightarrow \ldots \rightarrow r_3 \rightarrow p. \tag{3}$$

Thus, by the Law of transitivity, Theorem 14 of Chapter 2, we see that $p \rightarrow q$ is a tautology. But, this is the same as saying $p \Rightarrow q$.

Example 4: Show that the square of every even number is also an even number.

Solution
To prove this assertion, it is well to first rewrite it as:

$$p \text{ is an even number} \Rightarrow p^2 \text{ is an even number.} \quad (4)$$

Now, let us prove (4) using the method of direct proof:

$$p \text{ is an even number} \Rightarrow p = 2k, \quad \text{for some whole number } k.$$

$$\Rightarrow p^2 = (2k)^2$$
$$\Rightarrow p^2 = 4k^2$$
$$\Rightarrow p^2 = 2(2k^2)$$
$$\Rightarrow p^2 = 2n, \quad \text{where } n = 2k^2 \text{ is a whole number.}$$
$$\Rightarrow p^2 \text{ is an even number.}$$ ♥

THE METHOD OF INDIRECT PROOFS

The indirect proof method is based on the *law of contraposition* seen in Theorem 20 of Chapter 2. What this method says is that to prove

$$p \Rightarrow q, \quad (5)$$

it suffices to prove its contrapositive

$$\sim q \Rightarrow \sim p. \quad (6)$$

Indeed, establishing (6) is the same as showing that $\sim q \rightarrow \sim p$ is a tautology, which, by the Law of contraposition, is equivalent to proving that $p \rightarrow q$ is a tautology, which in turn is the same as establishing (5).

Therefore, whenever, we are asked to show (5), we may prove (6) instead, provided the latter is easier to establish. The method of indirect proofs is sometimes called **proof by contraposition**.

Example 5: Show that if the square of a whole number is even, the number too is even.

Solution

To prove this assertion, it is well to first rewrite it as:

$$p^2 \text{ is an even number} \Rightarrow p \text{ is an even number.} \qquad (7)$$

It turns out that it is easier to prove the contrapositive of (7), namely:

$$p \text{ is } \textbf{not} \text{ an even number} \Rightarrow p^2 \text{ is } \textbf{not} \text{ an even number.} \qquad (8)$$

So, let us establish the contrapositive instead:

$$\begin{aligned}
p \text{ is } \textbf{not} \text{ an even } &\Rightarrow p \text{ is an } \textbf{odd} \\
&\Rightarrow p = 2k + 1 \quad \text{for some whole number } k. \\
&\Rightarrow p^2 = (2k+1)^2 \\
&\Rightarrow p^2 = 4k^2 + 4k + 1 \\
&\Rightarrow p^2 = 2(2k^2 + 2k) + 1 \\
&\Rightarrow p^2 = 2n + 1, \quad \text{where } n \text{ is the whole number } 2k^2 + 2k. \\
&\Rightarrow p^2 \text{ is an } \textbf{odd} \text{ number.} \\
&\Rightarrow p^2 \text{ is } \textbf{not} \text{ an even number.}
\end{aligned}$$

Thus, by the method of direct proof, we have proved (8), the contrapositive of (7). But this in turn proves (7). Therefore, what we have actually done here is to combine two methods; namely, the method of indirect proof [proof by contraposition] with the method of direct proof. ♥

THE METHOD OF PROOF BY CONTRADICTION

This method may be used to prove a theorem that is given in the form of a tautology:

$$\boldsymbol{P} \qquad (9)$$

The method of proof by contradiction says that to prove (9), it is sufficient to show

$$\boldsymbol{\sim P \Rightarrow CONT}. \tag{10}$$

Indeed, showing that (10) is true is the same as proving that the conditional $\boldsymbol{\sim p \rightarrow CONT}$ is a tautology, which by virtue of the Law of contradiction, Theorem 22 of Chapter 2, is in turn the same as establishing that $\boldsymbol{P}$ is a tautology.

Example 6: Show that the number $\sqrt{2}$ is an irrational number.

Solution

It turns out this proposition can easily be proved by contradiction. So, let up assume the denial of the proposition is true, and let us explore the consequences of such an assumption. If the consequences are a contradiction, then we have established the proposition:

$\sqrt{2}$ is an **not** irrational $\Rightarrow \sqrt{2}$ is a **rational**

$$\Rightarrow \sqrt{2} = \frac{a}{b},$$

where a and b are whole umbers with no common factors. Indeed, we can always completely reduce the fraction a/b to satisfy this condition.

$$\Rightarrow 2 = \frac{a^2}{b^2}$$

$$\Rightarrow 2b^2 = a^2 \tag{11}$$

$$\Rightarrow a^2 \text{ is even}$$

$$\Rightarrow a \text{ is even,}$$

[see Example 5.]

This means that there is a whole number k such that:

$$a = 2k. \tag{12}$$

Now, putting (12) back into (11), we get:

$$2b^2 = 4k^2, \tag{13}$$

which after simplifying yields

$$b^2 = 2k^2 . \tag{14}$$

Hence, b^2 is even. But, this implies that b too is even. At this point we need not proceed any further with the proofing. For, we have arrived at a conclusion that both a and b are divisible by 2, which contradicts the assumption that these two whole numbers share no common factors. ♥

Remark: As mentioned earlier, sometimes the theorem we wish to prove may be stated in the form of an implication: $p \Rightarrow q$. To prove a theorem stated in this form, using the method of contradiction, requires that one establishes that the denial of $p \Rightarrow q$ leads to a contradiction. [We remind the reader that $p \Rightarrow q$ stands for "$p \rightarrow q$ is a tautology". Hence, the denial of $p \Rightarrow q$ is "$p \rightarrow q$ is not a tautology", which is the same as saying that we may have p **true**, while q is **false**.]

THE METHOD OF PROOF BY EXHAUSTION OF ALL CASES

To prove a theorem of the form:

$$p \Rightarrow q \tag{15}$$

using this method, it suffices to come up with some proposition r such that all three of these implications hold true:

$$p \Rightarrow (r \vee \sim r), \quad r \Rightarrow q, \quad \text{and} \quad \sim r \Rightarrow q \tag{16}$$

Indeed, assume we are able to do so. Then, we have the following tautologies:

$$p \rightarrow (r \vee \sim r), \quad r \rightarrow q, \quad \text{and} \quad \sim r \rightarrow q \tag{17}$$

Hence, by the Law of exhaustion, Theorem 15 of Chapter 2, $p \rightarrow q$ too is a tautology, which establishes (15).

Example 7: Show that the product of any two consecutive whole numbers is an even number.

Solution
Let n and $n+1$ be any two consecutive whole numbers. Then we have two cases, and these cases are **exhaustive**[3]:

- *Case 1.* n is even :

n is even $\Rightarrow n = 2k$, for some whole number k.

$\Rightarrow n(n+1) = 2k(2k+1)$

$\Rightarrow n(n+1) = 2(2k^2+k)$

$\Rightarrow n(n+1) = 2p$, where $p = 2k^2+k$

$\Rightarrow n(n+1)$ is even.

- *Case 2.* n is odd :

n is odd $\Rightarrow n = 2k+1$, for some whole number k.

$\Rightarrow n(n+1) = (2k+1)(2k+2)$

$\Rightarrow n(n+1) = 2(2k+1)(k+1)$

$\Rightarrow n(n+1) = 2p$, where $p = (2k+1)(k+1)$

$\Rightarrow n(n+1)$ is even.

Since, either case leads to the fact that $n(n+1)$ is even, we conclude that the product of any two consecutive whole numbers n and $n+1$ is even. ♥

[3] The modifier "exhaustive" is used here to mean that these cases do cover all possible situations

THE METHOD OF PROOF BY CONSTRUCTION

This method of proof is most suited to proving **existence theorems**; that is to say, theorems which claim the existence of some mathematical objects. Proving such a theorem by the construction method consists in simply exhibiting or [as the term suggests] constructing the object in question, and then verifying that this object does indeed satisfy all the defining conditions or properties stated in the theorem.

Example 8: Show that between any two nonnegative distinct rational numbers, there exists at least one irrational number r.

Solution

Lets a and b be rational numbers such that $0 \leq a < b$. We will construct our candidate irrational number r as follows: We try

$$r = a + \frac{b - a}{\sqrt{2}} \tag{18}$$

Next, we have to verify that r lies indeed between a and b, and that r is indeed irrational.

- Note that $0 < \frac{b-a}{\sqrt{2}} < b - a$. Therefore, adding a to all three sides, we obtain:

  $$a < a + \frac{b-a}{\sqrt{2}} < b \tag{19}$$

  which proves

  $$a < r < b \tag{20}$$

  In other words, we have just shown that r lies strictly between a and b.

- To prove the irrationality of r, we shall argue by contradiction:

Suppose r is not irrational. Then r is rational, and from (18) we would have

$$\sqrt{2} = \frac{b-a}{r-a}.$$

However, from this we infer that $\sqrt{2}$, being the ratio of rational numbers, must itself be rational, which is in contradiction with our finding in Example 6.

This completes the entire proof. 

3.3. CONCLUSION

In the foregoing sections we discussed and illustrated various strategies of proof. The list of proof methods presented here is however not an exhaustive one, but one that contains six of the methods most commonly used in mathematics. There are other important proof strategies: The so-called **methods of mathematical induction**. We will discuss these methods in Chapter 12.

In general, proofs encountered in the mathematical literature [and in the subsequent chapters for that matter] are more involved and more sophisticated than the examples presented in this chapter. These proofs often use, instead of one single method, a subtle combination of several proof strategies. We will have numerous opportunities to construct such proofs as we progress in our study. Hence the examples presented in this chapter are aimed at helping you to recognize the internal structures of many of the proof methods used by professional mathematicians.

A few lines of remark are in order at this point. As mentioned in the beginning of this chapter, some theorems may be stated in the form:

$$p \Leftrightarrow q. \tag{21}$$

Though it may seem that we have avoided this form, this is not quite the case. Indeed, thanks to Theorem 27 of Chapter 2, we know that (21) is simply the combination of a **forward** implication with its **converse** implication:

$$p \Rightarrow q \quad \text{and} \quad q \Rightarrow p.$$

Hence, to establish (21), one should prove both the forward implication and its converse. One may conduct both proofs separately or simultaneously. In the case of a simultaneous attack [where at every stage, we make sure both implications hold], it suffices that we establish a chain of equivalences linking the proposition p to the proposition q:

$$p \Leftrightarrow r_1 \Leftrightarrow r_2 \Leftrightarrow r_3 \Leftrightarrow \ldots \Leftrightarrow r_{n-1} \Leftrightarrow q.$$

PROBLEMS

1. In this chapter, we defined a theorem as one of three things:
   1.1. A tautology $\boldsymbol{P}$,
   1.2. An implication $\boldsymbol{P} \Rightarrow \boldsymbol{Q}$,
   1.3. Or an equivalence $\boldsymbol{P} \Leftrightarrow \boldsymbol{Q}$.

   Show that a theorem may be defined simply as an implication:

   $$\boldsymbol{P} \Rightarrow \boldsymbol{Q}.$$

2. Show that between any two rational numbers there is another rational number different from the previous two. [Name all your proof strategies for this problem.]
3. For now, we define a nonzero real number as the measure of a line segment. Using geometrical arguments base on similar triangles, show that there is a real number r such that $r^2 = n$. [Name all your proof strategies for this problem.]
4. Suppose r is a rational number and s an irrational number. Show that
   4.1. Their sum is $r + s$ is irrational
   4.2. Their product rs is irrational, if $r \neq 0$.

[Name all your proof strategies for this problem.]

5. * Use the method of indirect proof to show that

$$x^2 + xy + y^2 = 0 \Rightarrow [\, x = 0 \ \text{and} \ y = 0 \,]$$

6. Let p, q and r be any three propositions. Then show that the following properties are true.

6. 1. Reflexivity for equivalence:

$p \Leftrightarrow p$.

6. 2. Symmetry for equivalence:

If $p \Leftrightarrow q$, then $q \Leftrightarrow p$.

6. 3. Transitivity for equivalence:

If $p \Leftrightarrow q$ and $q \Leftrightarrow r$, then $p \Leftrightarrow r$.

6. 4. Substitution:

If $p \Leftrightarrow q$, then p may be replaced by q (or q by p) in any propositional expression, without changing the truth values of that propositional expression.

"Practise yourself, for heaven's sake, in little things; and thence proceed to greater."
Epictetus (Discourse IV. I)

CHAPTER 4

SETS

4.1. SET DEFINITION

In everyday life we deal with groups of objects. For example, when we use the term USA, we mean the group or collection of all 50 states that form the United States of America. Likewise, the word **class** often stands for a collection of students.

Definition 1: The term **set** is used in mathematics to refer to any **well-defined** collection of objects. The objects that make up a given set are called its **elements** or **members**.

Remarks 2: It is important to understand what we mean by the term **well-defined**. A collection is said to be well-defined, if it satisfies these three conditions:

1. Given any object whatsoever, the question "Does that object belong to the collection?" must have an answer, and only one of these two answers: either "yes" or "no"[4].

[4] It is not necessary that one personally has the knowledge required to decide which one of the two answers is correct. All that matters is that we know that one and only one of these answers applies.

2. The elements of a well-defined collection must be distinguishable from one another. [However, the order in which these elements are taken has no significance.]

3. No well-defined collection is a member of itself. ■

Example 3:
Which one of the following collections is a set?

A : The collection of the first five letters of the English Alphabet.

B : The collection of integer solutions to the equation: $x^2 - 25 = 0$.

C : The collection of beautiful people.

Solution:
Since the members of A are exactly a, b, c, d, and e, given any object, we know for sure whether or not that objects belongs to A. Hence A is a well-defined collection and a set for that reason. B too is a set, since the equation $x^2 - 25 = 0$ has exactly the two solutions: -5 and 5. As for the collection C, membership or nonmembership of any given person to it would depend pretty much on the taste of the individual entrusted with making the decision as to who is beautiful. Hence C is not well-defined, and not a set for that matter. ♥

To convey symbolically that an object x is a member of a set S, we shall write:

$$x \in S.$$

On the other hand, the negation $\sim (x \in S)$, which stands for the sentence "x is not a member of S" is usually written as:

$$x \notin S.$$

Example 4:
Let $\mathbb{N}$ stand for the set of counting numbers $1, 2, 3, \ldots$, also called **natural numbers**. Then which ones of the following statements are true?

1. $10 \in \mathbb{N}$
2. John $\in \mathbb{N}$
3. $-10 \in \mathbb{N}$
4. Mississippi $\in \mathbb{N}$

Solution:
Of these four statements, only the first is true. The second is not true, since John is not a number. The third is also not true, since -10, being a negative number, cannot be a counting number. The fourth statement is obviously false, as Mississippi is clearly not a number but a state. ♥

4.2. SET DESCRIPTION

In mathematics, a set may be described in one of three ways: (1) by giving a **word description** of the set, (2) by **listing** the members of the set, or (3) by providing a **defining property** of the set.

WORD DESCRIPTION METHOD

As its name suggests, this method simply consists in giving an accurate **verbal** or **word description** of the set in question

Example 5:
Here are a few instances of word descriptions of sets:

1. The set of all even numbers greater than 3, but less than 10.
2. The set of all women ever voted into the US congress.
3. The set of all graduates from Harvard University.
4. The set of all odd numbers.
5. The set of all squares.
6. The set of all fractions.

Note that in each case above, the word description provides a clear criterion for membership in the set. Therefore, these are all well-defined sets. ♥

LISTING OR ROSTER METHOD

Another way of describing a set is to list all of its elements, and only its elements, within a pair of curly brackets { }, and to separate these elements from one another by commas. This method is called the **listing** or **roster method**.

Example 6:

1. Let us denote by V the set of vowels found in the sentence "Bill Clinton was the forty second American president." Then we should write:

$$V = \{a, e, i, o\},$$

and read "V is the set made up of a, e, i, and o." Note that every member of set V is listed only once; **this is required in order to comply with the convention that the elements of a set must be distinct from one another**.

2. Some sets that contain a large or infinite number of elements, can still be described by the roster method provided the method is slightly modified. For instance, the set E of even numbers, and the set F of whole numbers less than 1000 may be represented as follows:

$$E = \{2, 4, 6, \ldots\}$$

and

$$F = \{0, 1, 2, \ldots, 1000\},$$

where the three dots stand for the term "**and so on**," and suggest that the listing continues according to the pattern exhibited by the first few elements already listed. ♥

DEFINING PROPERTY METHOD

The third method for describing sets is the **defining property method** or **set-builder notation**. This method is particularly convenient when dealing with infinite sets.

This is how we use the method to describe a set S :

1. We first find a property P which all members of the set S satisfy, but nonmembers of S do not satisfy. Such a property, if found, is called a **defining property** of the set S .

2. Then we write

$$S=\{x: \ x \text{ has the property } P\}, \qquad (1)$$

and call it the **set-builder** notation for set S .

The colon ":" in (1) stands for the phrase "**such that**". Also, the pair of curly brackets "{ }", within which everything is enclosed, is read "The set of all". Thus, the set-builder notation (1) may be read as " S is the set of all (objects) x , such that x has the property P ."

For convenience, (1) is often written in the more compact form:

$$S=\{ x: \ P(x) \}, \qquad (2)$$

and read simply as " S is the set of all x such that $P(x)$ is true."

Before looking at some examples it is worth emphasizing the following obvious result:

Theorem 7:

$$z \in \{ x: P(x)\} \Leftrightarrow P(z). \qquad (3)$$

In other words, an object z that belongs to the set $\{ x: P(x)\}$ must satisfy the defining property P ; and conversely, any object z that satisfies property P must belong to the set $\{ x: P(x)\}$.

Proof: This theorem is a straightforward consequence of the set-builder notation. ♣

Example 8: Describe the following sets by using the set-builder notation.

1. $A = \{2, 4, 6, 8\}$,
2. B: the set of all US presidents,
3. $C = \{1, 3, 5, \ldots\}$.

Solution:

1. Let us denote by $\mathbb{N}$ the set of natural numbers. Since A is made up of those natural numbers that are divisible by 2 and smaller than 9, we may write:

$$A = \{ y : \ (y \in \mathbb{N}),\ (y \text{ is divisible by } 2) \text{ and } (y \leq 9) \}.$$

2. For B, letting H be the set of all humans, we easily see that:

$$B = \{ z : \ (z \in H) \text{ and } (z \text{ is a US president}) \}.$$

3. Now, C being the set of all odd numbers, we may write:

$$C = \{ x : \ (x \in \mathbb{N}) \text{ and } (x \text{ is odd}) \}.$$ ♥

Example 9: Describe the following set using the listing method.

$$A = \{ w : \ (w \in \mathbb{N}) \text{ and } (w + 10 = 0) \}.$$

Solution:

Note that set A is described here by the set-builder notation. Thus, for any object x to be member of A, that object x must satisfy the defining property:

$$x \in \mathbb{N}, \quad \text{and} \quad x + 10 = 0.$$

That is to say, in order for x to be in set A, it must be a counting number which added to 10 yields 0. However, no such counting number exists. We express this situation by saying that set A contains no elements. A set such as A, that contains no elements whatsoever is said to be **empty**. ♥

Definition 10: The **empty** set is the set that contains no elements. It is symbolized by the Norwegian letter $\varnothing$.

4.3. SUBSETS

Sometimes, it does happen that all the elements of one set A are also elements of another set B. For instance, if we call A the set of people living in the state of Illinois, and B the set of people living in the USA, then clearly, every element of A is also an element of set B. In a situation such as this, the set A is said to be a **subset** of set B, whereas set B is called a **superset** of set A.

Definition 11: Let A and B be two sets. We say that A is a **subset** of B or that A is **contained** in B, and we write $A \subset B$, if every element of A is also an element of B. Sometimes, we say that B is a **superset** of A to mean exactly that A is a subset of B, in which case we write $B \supset A$.

Remark 12: Hence, using logically symbolism

$$A \subset B \quad \Leftrightarrow \quad (x \in A \Rightarrow x \in B).$$

To put it in a less formal way, a subset of a set B is the set of elements left after one has deleted from B all, some, or none of its elements. ■

Example 13:

If $A = \{a,\ s\}$, $B = \{a,\ e,\ s\}$, and $C = \{e,\ s\}$, then which ones of the following statements are true?

1. $A \subset B$.   2. $B \subset C$.   3. $C \subset B$.

Solution:
The first statement is true, since every element of A is also an element of B. The second statement is not true, since not every element of B is an element of C. The third statement is true; indeed every element of C is also a member of B. Hence, of these statements, only the second one is false ♥

We now can prove the following theorem.

Theorem 14: For every set S, $S \subset S$ and $\varnothing \subset S$.

Proof:

For every set S, we do have $S \subset S$, since every element of S is indeed an element of S. As for the proof of $\varnothing \subset S$, we gave one in Example 3 of Chapter 3. ♣

From this theorem, we see that any non-empty set A has at least two distinct subsets; namely, the empty set $\varnothing$ and the set A itself. These two subsets are called **trivial** subsets of A. And any subset of A other than the set A itself is called a **proper** subset of A.

The collection of all subsets of any set A is called the **power set** of A and will be denoted by 2^A. The following theorem gives us the total numbers of elements in a power set 2^A.

Theorem 15: Let A be a set containing n elements, where n is a natural number. Then 2^A contains 2^n elements.

Proof [by exhaustion of all cases]:

Let A be a set containing n elements. Then we know that any subset S of A should fall into exactly one of these $n+1$ cases:

Case 0. $S = \varnothing$: In this case, using elementary counting[5] principles, we immediately see that the number of distinct subsets S, of A, with no elements at all is C_n^0 .

Case 1. S has exactly 1 element: In this case too, using elementary counting principles, we see that the number of distinct subsets S, of A, with exactly one element is C_n^1 .

Case 2. S has exactly 2 elements: The number of distinct subsets S, of A, containing exactly two elements is C_n^2 .

....

....

....

Case n. S has exactly n elements: The number of distinct subsets S, of A, having exactly n elements is C_n^n .

Thus, the total number of distinct subsets S of A is:

$$\begin{aligned} C_n^0 + C_n^1 + C_n^2 + \ . \ . \ . + C_n^2 &= C_n^0 1^0 1^{n-0} + C_n^1 1^1 1^{n-1} + C_n^2 1^2 1^{n-2} \\ &\quad + \ldots + C_n^{n-1} 1^{n-1} 1^1 + C_n^n 1^n 1^0 \\ &= \sum_{i=0}^{n} C_n^i 1^i 1^{n-i} \\ &= (1+1)^n, \quad \text{by Newton's Binomial Theorem}^6. \\ &= 2^n, \end{aligned}$$

which completes the proof. ♣

---

[5] See your high-school or college algebra notes of the theory on counting.

[6] Refer to your high-school or college algebra notes.

4.4. EQUAL SETS

If you can remember, we said that the specific order in which the elements of a set are listed is of no importance. Thus, the following lists $\{a, e, i, o, u\}$ and $\{o, a, i, e, u\}$ denote one and the same set.

Definition 16: Let A and B be any two sets. We say that A and B are **equal**, written as $A = B$, if the two sets are made up of exactly the same objects.

Our next theorem is a direct consequence of the preceding definitions

Theorem 17: For any two sets A and B:

$$(A = B) \Leftrightarrow (A \subset B \text{ and } B \subset A) \tag{4}$$

Proof:
To show this theorem, we need to prove two things:

- **Proof of** $(A = B) \Rightarrow (A \subset B$ **and** $B \subset A)$.
  We shall use the method of direct proofs. Suppose $A = B$. This means that the sets A and B have exactly the same elements. Thus, every element of A is also an element of B, and vice versa, every element of B is also an element of A Therefore, $A \subset B$ and $B \subset A$.

- **Proof of** $(A \subset B$ **and** $B \subset A) \Rightarrow (A = B)$.
  We shall use the method of proof by contradiction. Assume that $(A \subset B$ and $B \subset A) \Rightarrow (A = B)$ is not true. Then, there are at least two sets A and B such that

$$A \subset B, \tag{5}$$

$$B \subset A \tag{6}$$

and yet

$$A \neq B. \tag{7}$$

But then (7) would means that the two sets A and B do not have the same elements; which in turn would contradict at least one of the two propositions (5) or (6). Hence, we have completed the proof by contradiction. ♣

Example 18: For any two sets A and B, prove that

$$(A \neq B) \;\Leftrightarrow\; [\sim(A \subset B) \text{ or } \sim(B \subset A)] \tag{8}$$

Solution:
From Theorem 17, we know that

$$(A = B) \;\Leftrightarrow\; [\,(A \subset B) \text{ and } (B \subset A)] \tag{9}$$

Therefore, by contraposition,

$$\sim(A = B) \;\Leftrightarrow\; \sim[\,(A \subset B) \text{ and } (B \subset A)]\,. \tag{10}$$

That is

$$(A \neq B) \;\Leftrightarrow\; \sim[\,(A \subset B) \text{ and } (B \subset A)]\,, \tag{11}$$

which, by one of DeMorgan's Laws of logic, yields

$$(A \neq B) \;\Leftrightarrow\; [\sim(A \subset B) \text{ or } \sim(B \subset A)]\,,$$

as we wished to prove. ♥

4. 5. TRUTH SETS

UNIVERSAL SET

It is often convenient to assume that all sets involved in a discussion are subsets of one given "large" set U. In that case, the set U is called the **universal** set or the **universe of discourse**. Thus, for example, in a school there will be a set of mathematics professors, a set of English professors, a set of science professors, etc. And all these different sets would be subsets of the universal set U of all professors of the school.

Definition 19: The **universal** set U for a particular discussion is the set made up of all elements in that discussion or situation.

Remark 20: For every problem that we discuss, we must clearly be aware of the universal set, which must remain fixed throughout the discussion. Thus, in the remainder of this text, when the universal set U is not explicitly mentioned, we will assume that it is the set $\mathbb{R}$ of real numbers. ■

VARIABLE

You are certainly aware that the area a circle is given by the formula $3.14r^2$, where r is the radius of the circle. Hence, if asked what is the numerical value of the area of a circle? you probably will answer by saying that it depends. Indeed, as the formula shows, the area of a circle depends on the value taken by the radius r.

In fact, the value of r varies; r may assume any non-negative value. A letter or symbol such as r which may take any value from a given set is called a variable.

Definition 21: In algebra, a **variable** is a letter or symbol which may be replaced by any element of a given universal set U.

Variables are usually named after letters at the end of the English alphabet: x, y and z. However, to make it convenient to remember what quantity a variable stands for, we often choose to symbolize the variable by the first letter in the name of that quantity. Thus, for example, we used r for the radius of a circle.

In contrast to a variable, there is also the idea of a constant.

Definition 22: A **constant** is a symbol that stands for a fixed element of a universal set U.

OPEN SENTENCES

We learned in Chapter 1 that the sentence

$$x^2 - 25 = 0, \quad \textit{with} \quad x \in \mathbb{R}, \tag{12}$$

is not a proposition since it is not possible to say whether it is true or false without first specifying a fixed replacement for the variable x from the universal set $\mathbb{R}$. Indeed, (12) is true for $x = 5$ or $x = -5$, but false for any other value of x from set $\mathbb{R}$. A sentence such as $x^2 - 25 = 0$, which is not a proposition until and unless we have specified an actual replacement value for its variable x, is called an **open sentence** or a **predicate**.

> **Definition 23:** An **open sentence** or **predicate** is any sentence containing a variable, which becomes a proposition whenever the variable is assigned a value from its universal set *U*.

TRUTH SET OF AN OPEN SENTENCE

Let us consider once again the open sentence:

$$x^2 - 25 = 0, \qquad x \in \mathbb{R}.$$

Then, the set

$$\{x \in \mathbb{R} : \quad x^2 - 25 = 0\},$$

made up of all $x \in \mathbb{R}$ for which the open sentence holds true, is called the **truth set** of this open sentence.

Example 24:

Show that the truth set of $x^2 - 25 = 0,\ x \in \mathbb{R}$ is $\{-5,\ 5\}$.

Solution:

To show this is the case, we have to prove that

$$\{x \in \mathbb{R} : \quad x^2 - 25 = 0\} = \{-5,\ 5\}, \tag{13}$$

which, by Theorem 17, is the same as showing that both of these propositions hold:

$$\{x \in \mathbb{R} : \quad x^2 - 25 = 0\} \subset \{-5, 5\} \tag{14}$$

and

$$\{-5, 5\} \subset \{x \in \mathbb{R} : \quad x^2 - 25 = 0\} \tag{15}$$

- **Proof of** $\{x \in \mathbb{R} : \quad x^2 - 25 = 0\} \subset \{-5, 5\}$**:** It suffices to show, by direct proof techniques, that if an element belongs to the set on the right, then it also belongs to the set on the left.

$$\begin{aligned} p \in \{x \in \mathbb{R} : \quad x^2 - 25 = 0\} &\Rightarrow p \text{ satisfies } x^2 - 25 = 0 \\ &\Rightarrow p^2 - 25 = 0 \\ &\Rightarrow (p-5)(p+5) = 0 \\ &\Rightarrow \begin{cases} p - 5 = 0, \\ \text{or} \\ p + 5 = 0 \end{cases} \\ &\Rightarrow \begin{cases} p = 5, \\ \text{or} \\ p = -5 \end{cases} \\ &\Rightarrow p \in \{-5, 5\}. \end{aligned}$$

Thus, $p \in \{x \in \mathbb{R} : x^2 - 25 = 0\} \Rightarrow p \in \{-5, 5\}$.

Hence, we have shown that

$$\{x \in \mathbb{R} : \quad x^2 - 25 = 0\} \subset \{-5, 5\}. \tag{16}$$

- **Proof of** $\{-5, 5\} \subset \{x \in \mathbb{R} : \quad x^2 - 25 = 0\}$**:** We will combine here the technique of direct proofs with that of **proofs by exhaustion to show that** if an element belongs to the set on the right, then it also belongs to the set on the left.

$$p \in \{-5,\ 5\} \Rightarrow \begin{cases} p = 5, \\ \text{or} \\ p = -5 \end{cases}$$

$$\Rightarrow \begin{cases} p^2 = (5)^2, \\ \text{or} \\ p^2 = (-5)^2 \end{cases}$$

$$\Rightarrow \begin{cases} p^2 = 25, \\ \text{or} \\ p^2 = 25 \end{cases}$$

$$\Rightarrow \begin{cases} p^2 - 25 = 0, \\ \text{or} \\ p^2 - 25 = 0 \end{cases}$$

$$\Rightarrow\ p^2 - 25 = 0$$

which, by Theorem 7, implies that

$$p \in \{x \in \mathbb{R}: \quad x^2 - 25 = 0\}.$$

Thus, $p \in \{-5,\ 5\} \Rightarrow p \in \{x \in \mathbb{R}: \quad x^2 - 25 = 0\}$.

Hence, we have arrived at the fact that

$$\{-5,\ 5\} \subset \{x \in \mathbb{R}: \quad x^2 - 25 = 0\}. \qquad (17)$$

Finally, combining (16) and (17) we get

$$\{x \in \mathbb{R}: \quad x^2 - 16 = 0\} = \{-4,\ 4\}. \qquad \heartsuit$$

In general, we will represent an open sentence about a single variable x by $P(x)$, $Q(x)$, or $R(x)$.

Definition 25: Let $P(x)$ be an open sentence about a single variable x. Then the **truth set** of $P(x)$ is the set

$$\{\, x \in U : \; P(x) \,\}$$

of all elements $x \in U$ for which the open sentence $P(x)$ is true.

From the forgoing definition we automatically have

Corollary 26:

$$a \in \left\{\, x \in U : \; P(x) \right\} \;\Leftrightarrow\; \left[\, a \in U, \textit{ and } P(a) \textit{ is true} \right] \qquad (18)$$

Proof: The corollary is a direct consequence of Definition 25 and the set-builder notation. ♣

Remark 27: In the remainder of this text, instead of the profuse and therefore cumbersome notation

$$a \in U, \quad \textit{and } P(a) \textit{ is true}$$

we will simply write

$$(a \in U) \wedge P(a)$$

to signify the same thing. ■

Remark 28: At times, we run into open sentences $P(x)$ that are true for all values x of their universal set U. Thus, clearly in such cases, we will have:

$$\left\{\, x \in U : \; P(x) \,\right\} = U \,.$$

Any open sentence $P(x)$ which is verified by all values of x in a given universal set U is said to be an **absolutely true open sentence** on that universal set. ■

Example 29: Let us choose $U = \mathbb{R}$, the set of real numbers. Explain why we have:

1. $\{x \in \mathbb{R} : \; x^2 - 1 = (x+1)(x-1)\} = \mathbb{R}$.
2. $\{x \in \mathbb{R} : \; x^2 + 1 > 0\} = \mathbb{R}$.

Solution:

1. Note that the open sentence $x^2 - 1 = (x-1)(x+1)$ is an equation which is true for all values of x in $\mathbb{R}$. Hence, by Remark 28, we must have:

$$\{x \in \mathbb{R} : \; x^2 - 1 = (x+1)(x-1)\} = \mathbb{R}.$$

An absolutely true equation, such as

$$x^2 - 1 = (x-1)(x+1)$$

is customarily referred to as an **identity**.

2. Similarly, it is not difficult to see that the inequality $x^2 + 1 > 0$ remains true for all values of x in $\mathbb{R}$. Hence, it is an **absolutely true inequality**. Therefore, we do have:

$$\{x \in \mathbb{R} : \; x^2 + 1 > 0\} = \mathbb{R}. \; ♥$$

Remark 30: There are other times too when we encounter open sentences $P(x)$ that are false for all values x in the universal set U. Thus, in these cases, we must have:

$$\{\, x \in U : \; P(x) \,\} = \varnothing.$$

A open sentence $P(x)$ which is false for any value x in a given universal set U is said to be **absolutely false** on that universal set.

Example 31: Let us choose $U = \mathbb{R}$, the set of real numbers. Explain why we have:

1. $\{x \in \mathbb{R} : \; x^2 = -1\} = \varnothing$.
2. $\left\{x \in \mathbb{R} : \; \frac{x}{x} < 0\right\} = \varnothing$.

Solution:

1. Note that the open sentence $x^2 = -1$ is an equation which is satisfied by no real number. Hence $x^2 = -1$ is an absolutely false equation. Thus, by the foregoing remark:

$$\{x \in \mathbb{R} : x^2 = -1\} = \varnothing .$$

2. It is also easy to see that the inequality $\frac{x}{x} < 0$ is satisfied by no real number. Indeed, $\frac{x}{x}$ is either undefined, for $x = 0$, or otherwise equal to 1. Thus,

$$\left\{x \in \mathbb{R} : \frac{x}{x} < 0\right\} = \varnothing . \ ♥$$

PROBLEMS

1. Describe these sets using the Listing Method
   - 1.1. $S = \{x : x \in \mathbb{R} \text{ and } x^2 - 6x + 5 = 0\}$
   - 1.2. $S = \{x : x \text{ is an odd number, and } 3 \leq x \leq 12\}$
   - 1.3. $S = \{x : x \text{ is a divisor of } 24 \text{ and } 30\}$
2. Describe these sets using the Set-Builder notation
   - 2.1. $S = \{1, 2, 3, 4, 5\}$
   - 2.2. S = The set of all real numbers greater than 10.
   - 2.3. S = The set of all real numbers less than 20.
   - 2.4. S = The set of all four-sided triangles.
   - 2.5. S = The set of male hens.
3. **Russell's Paradox:** Let S be the collection defined by

$S = \{x : x \notin x\}$.

3.1. Show that the proposition $S \notin S$ leads to a contradiction.

3.2. Show that $S \in S$ also leads to a contradiction.

3.3. Is S a well-defined collection?

3.4. Is S a set?

4. To what sets are these sets equal to:

4.1. $S = \{x : x \in \mathbb{R} \text{ and } x^2 + 5 < 0\}$?

4.2. $S = \{x : x \in \mathbb{R} \text{ and } x^2 + 5 > 0\}$?

5. Let S be the collection all mammals

5.1. Is $S \in S$?

5.2. Is $S \notin S$?

5.3. Is S a well-defined collection?

5.4. Is S a set?

"Civilization advances by extending the number of important operations which we can perform without thinking about them"
Alfred North Whitehead

CHAPTER 5

OPERATIONS ON SETS

Now that we know what sets are, and how they may be described, we are will study some of the basic operations performed on them. The operations we will examine in this chapter are: **intersection**, **union**, **complement**, **difference**, and **Cartesian product**.

5.1. THE INTERSECTION OF TWO SETS

Definition 1: Let A and B be any two subsets of a universal set U. The **intersection** of A with B, denoted by $A \cap B$, is the set of all those elements of U that are common to A and B.

Remark 2: Clearly, from the foregoing definition,

$$A \cap B = \{x \in U : x \in A \textbf{ and } x \in B\}. \qquad (1)$$

In other words,

$$x \in A \cap B \iff x \in A \textbf{ and } x \in B. \blacksquare \qquad (2)$$

Example 3:

Let $A = \{2, 5, 7, 8, 10, 12\}$, $B = \{3, 5, 9, 10, 13\}$,

and $C = \{2, 12\}$.

Describe the following intersections

1. $B \cap C$.   2. $A \cap B$.

Solution:
1. $B \cap C = \varnothing$, since sets B and C have no elements in common. Two sets, such as B and C, whose intersection is the empty set are will be said to be **disjoint**.

2. $A \cap B = \{5, 10\}$. Two sets, such as A and B, whose intersection is not empty are said to be **overlapping**. ♥

Example 4:

Let U be our universal set, $\varnothing$ the empty set, and A any subset of U. Then, show that:

1. $A \cap \varnothing = \varnothing$.   2. $A \cap U = A$.

Solution:
1. Since $\varnothing$ is empty, there is no elements shared by both set $\varnothing$ and set A. Thus, $A \cap \varnothing = \varnothing$.

2. Since U is the universal set, any set A is a subset of U. Therefore, every element of A is also an element of U. Thus $A \cap U = A$. ♥

We gather these results in the following theorem

Theorem 5: If U is the universal set, $\varnothing$ the empty set, and A any subset of U, then:

$$A \cap \varnothing = \varnothing. \quad (3)$$

and

$$A \cap U = A. \quad (4)$$

In order to introduce our next theorem, we first need this definition:

Definition 6: Let $P(x)$ and $Q(x)$ be open sentences. The conjunction of $P(x)$ and $Q(x)$ is the open sentence " $P(x)$ **and** $Q(x)$," symbolized by $P(x) \wedge Q(x)$.

We are now ready to make this very important claim

Theorem 7: Let $P(x)$ and $Q(x)$ be open sentences about the variable $x \in U$. Then:

$$\{x \in U : P(x) \wedge Q(x)\} = \{x \in U : P(x)\} \cap \{x \in U : Q(x)\}. \qquad (5)$$

In other words, the truth set of the conjunction of two open sentences is the intersection of the truth sets of these open sentences.

Proof:

To prove (5) we need to show two things:

$$\{x \in U : P(x)\} \cap \{x \in U : Q(x)\} \subset \{x \in U : P(x) \wedge Q(x)\} \qquad (6)$$

and

$$\{x \in U : P(x) \wedge Q(x)\} \subset \{x \in U : P(x)\} \cap \{x \in U : Q(x)\} \qquad (7)$$

- **Direct proof of (6):** it is enough to show that every element of the set to the left of the inclusion sign, $\subset$, is also an element of the set on the right:

$$a \in \{x \in U : P(x) \wedge Q(x)\} \Rightarrow a \in U \text{, and } P(a) \wedge Q(a) \text{ is true}$$

$$\Rightarrow (a \in U) \wedge (P(a) \wedge Q(a))$$

$$\Rightarrow \begin{cases} (a \in U) \wedge (P(a)), \\ and \\ (a \in U) \wedge (Q(a)), \end{cases}$$

where we have applied the result of Problem 10 of Chapter 2

$$\Rightarrow \begin{cases} a \in \{\, x \in U :\ P(x) \,\}, \\ and \\ a \in \{\, x \in U :\ Q(x) \,\} \end{cases}$$

$$\Rightarrow a \in \{x \in U :\ P(x)\} \cap \{x \in U :\ Q(x)\}\,.$$

Hence, we have

$$a \in \{x \in U :\ P(x) \wedge Q(x)\}$$

implies

$$a \in \{x \in U :\ P(x)\} \cap \{x \in U :\ Q(x)\}.$$

Therefore:

$$\{x \in U :\ P(x) \wedge Q(x)\} \subset \{x \in U :\ P(x)\} \cap \{x \in U :\ Q(x)\} \qquad (8)$$

- **Direct proof of (7):** We know that

$$a \in \{x \in U :\ P(x)\} \cap \{x \in U :\ Q(x)\}$$

implies

$$a \in \{x \in U :\ P(x)\} \quad and \quad a \in \{x \in U :\ Q(x)\}\,. \qquad (9)$$

Also,

$$(9) \Rightarrow \begin{cases} (a \in U) \wedge (P(a)) \\ and \\ (a \in U) \wedge (Q(a)) \end{cases}$$

$$\Rightarrow (a \in U) \wedge (\, P(a) \wedge Q(a)\,)\text{, by Problem 10, Chapter 2}$$

$$\Rightarrow a \in \{x \in U :\ P(x) \wedge Q(x)\}\,.$$

Thus, we have shown that

$$a \in \{x \in U :\ P(x)\} \cap \{x \in U :\ Q(x)\}$$

implies

$$a \in \{x \in U :\ P(x) \wedge Q(x)\}.$$

Therefore:

$$\{x \in U :\ P(x)\} \cap \{x \in U :\ Q(x)\} \subset \{x \in U :\ P(x) \wedge Q(x)\}\,. \quad (10)$$

Finally, combining (8) and (10) we have:

$$\{x \in U : P(x)\} \cap \{x \in U : Q(x)\} = \{x \in U : P(x) \wedge Q(x)\}$$

as we wished to show . ♣

Example 8: Describe the following sets using interval notations:

1. $\{ x : x \geq 0 \ \textit{and} \ x \geq 1 \}$
2. $\{ x : x \leq 0 \ \textit{and} \ x \leq 1 \}$

Solution:

1. Using Theorem 7, we have:

$$\begin{aligned}\{ x : x \geq 0 \ \textit{and} \ x \geq 1 \} &= \{ x : x \geq 0 \} \cap \{ x : x \geq 1 \} \\ &= [\, 0, \ +\infty \,[\cap [\, 1, \ +\infty \,[\\ &= [\, 1, \ +\infty \,[\, .\end{aligned}$$

2. Similarly, using Theorem 7, we have:

$$\begin{aligned}\{ x : x \leq 0 \ \textit{and} \ x \leq 1 \} &= \{ x : x \leq 0 \} \cap \{ x : x \leq 1 \} \\ &=]-\infty, \ 0 \,] \cap]-\infty, \ 1 \,] \\ &=]-\infty, \ 0 \,] \, .\end{aligned}$$ ♥

5.2. THE UNION OF TWO SETS

Definition 9: Let A and B be subsets of a universal set U. The **union** of A with B, denoted by $A \cup B$, is the set of all those elements of U that belong to at least one of the two sets.

Remark 10: Thus, symbolically:

$$A \cup B = \{x \in U : x \in A \ \textbf{or} \ x \in B\} . \tag{11}$$

In other words,

$$x \in A \cup B \iff x \in A \textbf{ or } x \in B. \blacksquare \tag{12}$$

Example 11:

Let $A = \{2, 5, 7, 8, 10, 12\}$, $B = \{3, 5, 9, 10, 13\}$

Describe the union $A \cup B$.

Solution:

$A \cup B = \{2, 5, 7, 8, 10, 12, 3, 9, 13\}$. ♥

Definition 12: Let $P(x)$ and $Q(x)$ be open sentences. The disjunction of $P(x)$ and $Q(x)$ is the open sentence " $P(x)$ **or** $Q(x)$," symbolized by $P(x) \vee Q(x)$.

After this definition, we are ready to prove the following

Theorem 13: Let $P(x)$ and $Q(x)$ be open sentences about the variable $x \in U$. Then:

$$\{x \in U : P(x) \vee Q(x)\} = \{x \in U : P(x)\} \cup \{x \in U : Q(x)\}. \tag{13}$$

What this theorem says is that the truth set of the disjunction of two open sentences is the union of the truth sets of these open sentences.

Proof:

To prove (13) we must prove both of these:

$$\{x \in U : P(x)\} \cup \{x \in U : Q(x)\} \subset \{x \in U : P(x) \vee Q(x)\} \tag{14}$$

and

$$\{x \in U : P(x) \vee Q(x)\} \subset \{x \in U : P(x)\} \cup \{x \in U : Q(x)\} \tag{15}$$

- **Direct proof of (14) :**

We know that

$$a \in \{x \in U : \ P(x)\} \cup \{x \in U : \ Q(x)\}$$

implies

$$a \in \{x \in U : \ P(x)\} \quad or \quad a \in \{x \in U : \ Q(x)\}. \qquad (16)$$

Also

$$(16) \quad \Rightarrow \begin{cases} (a \in U) \wedge (P(a)) \\ or \\ (a \in U) \wedge (Q(a)) \end{cases}$$

$$\Rightarrow (a \in U) \wedge (P(a) \vee Q(a))$$

where we have used the result of Problem 11 of Chapter 2

$$\Rightarrow a \in \{x \in U : \ P(x) \vee Q(x)\}.$$

Thus,

$$a \in \{x \in U : \ P(x)\} \cup \{x \in U : \ Q(x)\}$$

implies

$$a \in \{x \in U : \ P(x) \vee Q(x)\}.$$

Therefore

$$\{x \in U : \ P(x)\} \cup \{x \in U : \ Q(x)\} \subset \{x \in U : \ P(x) \vee Q(x)\}. \quad (17)$$

- **Direct proof of (15) :** Here too, it is enough to show that every element of the set to the left of the inclusion sign, $\subset$, is also an element of the set on the right.

$$a \in \{x \in U : P(x) \vee Q(x)\} \Rightarrow (a \in U) \wedge (P(a) \vee Q(a))$$

$$\Rightarrow \begin{cases} (a \in U) \wedge (P(a)) \\ or \\ (a \in U) \wedge (Q(a)) \end{cases}$$

$$\Rightarrow \begin{cases} a \in \{x \in U : \ P(x)\} \\ \text{or} \\ a \in \{x \in U : \ Q(x)\} \end{cases}$$

$$\Rightarrow a \in \{x \in U : \ P(x)\} \cup \{x \in U : \ Q(x)\}.$$

Thus,

$$a \in \{x \in U : \ P(x) \vee Q(x)\}$$

implies

$$a \in \{x \in U : \ P(x)\} \cup \{x \in U : \ Q(x)\}.$$

This shows that

$$\{x \in U : \ P(x) \vee Q(x)\} \subset \{x \in U : \ P(x)\} \cup \{x \in U : \ Q(x)\}$$
(18)

Finally, combining (17) and (18) we have:

$$\{x \in U : \ P(x)\} \cup \{x \in U : \ Q(x)\} = \{x \in U : \ P(x) \vee Q(x)\},$$

which is what we wanted to show . ♣

Example 14: Find the domain of the function f defined by the formula:

$$f(x) = \sqrt{x^2 - x}.$$

Solution:

We know from algebra that the *default* domain D_f of a function f, in the case where D_f is not explicitly provided, is the totality of all real numbers x for which the image $f(x)$ is also a real number. Hence:

$$D_f = \{ x \in \mathbb{R} : \ f(x) \in \mathbb{R} \}$$

That is,

$$D_f = \left\{ x \in \mathbb{R} : \ \sqrt{x(x-1)} \in \mathbb{R} \right\}$$

$$= \{ x \in \mathbb{R} : \ x(x-1) \geq 0 \}. \qquad (19)$$

The inequality in (19) is due to the fact that square-roots are real, if and only if their radicands are non negative. Hence,

$$D_f = \{ x \in \mathbb{R} : \ x(x-1) \geq 0 \}$$
$$= \{ x \in \mathbb{R} : \ (x \geq 0 \ and \ x-1 \geq 0) \ \ or \ \ (x \leq 0 \ and \ x-1 \leq 0) \}$$
$$= \{ x \in \mathbb{R} : \ x \geq 0 \ and \ x-1 \geq 0 \} \cup \{ x \in \mathbb{R} : \ x \leq 0 \ and \ x-1 \leq 0 \}$$
$$= \{ x \in \mathbb{R} : \ x \geq 0 \ and \ x \geq 1 \} \cup \{ x \in \mathbb{R} : \ x \leq 0 \ and \ x \leq 1 \}.$$

Thus, from the results of Example 8, we have

$$D_f =]-\infty, \ 0] \cup [1, \ +\infty[. \ ♥$$

5.3. THE COMPLEMENT OF A SET

Sometimes, given a set A, one may want to know those elements of the universal set U that are not in A.

Definition 15: Let A be a subset of the universal set U. The **complement** of A, denoted by A' and read as " A **prime**," is the set of all those elements of U that do not belong to set A.

Example 16:
If $U = \{3, 5, 6, 12, 15, 31\}$, then what is the complement of the following set $A = \{6, 12, 31\}$?

Solution:
There are exactly three items of the universal set U that are not elements of set A; namely, 3, 5, and 15. Therefore,

$$A' = \{3, 5, 15\}, \ ♥$$

Remark 17: Thus, the above definition can be written as:

$$A' = \{ x \in U : \ x \notin A \} . \tag{20}$$

That is,

$$x \in A' \Leftrightarrow x \notin A . \ \blacksquare \tag{21}$$

Next, we will provide, using the set-builder notation, a simple characterization of complements of sets. But, to be able to do so, we will need the following definition.

Definition 18: Let $P(x)$ be an open sentence. The denial of $P(x)$ is the open sentence "**It is not the case that** $P(x)$". We will symbolize the denial of $P(x)$ by $\sim P(x)$, and simply read it as "**not** $P(x)$**.**"

Example 19:
Write the denial of the open sentence: $(x+2)(x-1)=0$.

Solution:
The denial of

$$(x+2)(x-1)=0$$

denoted by

$$\sim (x+2)(x-1)=0 \, ,$$

is

It is not the case that $(x+2)(x-1)=0$

or

It is false that $(x+2)(x-1)=0$,

which in turn is customarily written as

$$(x+2)(x-1) \neq 0 \, .$$

Thus, the negation of $(x+2)(x-1)=0$ is

$$(x+2)(x-1) \neq 0 \, . \ \heartsuit$$

We are now in the position to state our claim:

Theorem 20: Let $P(x)$ be an open sentence about the variable $x \in U$. Then

$$\{x \in U: P(x)\}' = \{x \in U: \sim P(x)\}.$$

Proof:

To show the theorem, we must show two things:

$$\{x \in U: P(x)\}' \subset \{x \in U: \sim P(x)\}$$

and

$$\{x \in U: \sim P(x)\} \subset \{x \in U: P(x)\}'.$$

- **Direct proof of** $\{x \in U: P(x)\}' \subset \{x \in U: \sim P(x)\}$:

$$\begin{aligned} a \in \{x \in U: P(x)\}' &\Rightarrow a \in U, \textit{ and } a \notin \{x \in U: P(x)\} \\ &\Rightarrow a \in U, \textit{ and } P(a) \textit{ is not true} \\ &\Rightarrow a \in U, \textit{ and } \sim P(a) \textit{ is true} \\ &\Rightarrow a \in \{x \in U: \sim P(x)\} \end{aligned}$$

Thus, $a \in \{x \in U: P(x)\}' \Rightarrow a \in \{x \in U: \sim P(x)\}$.

Hence, we have:

$$\{x \in U: P(x)\}' \subset \{x \in U: \sim P(x)\} \qquad (22)$$

- **Direct proof of** $\{x \in U: \sim P(x)\} \subset \{x \in U: P(x)\}'$:

$$\begin{aligned} a \in \{x \in U: \sim P(x)\} &\Rightarrow a \in U, \textit{ and } \sim P(a) \textit{ is true} \\ &\Rightarrow a \in U, \textit{ and } P(a) \textit{ is false} \\ &\Rightarrow a \text{ belongs } U, \\ &\qquad \text{but not to } \{x \in U: P(x)\} \\ &\Rightarrow a \in \{x \in U: P(x)\}'. \end{aligned}$$

Thus, $a \in \{x \in U: \sim P(x)\} \Rightarrow a \in \{x \in U: P(x)\}'$.

Hence, we have:

$$\{x \in U : \sim P(x)\} \subset \{x \in U : P(x)\}' \qquad (23)$$

Therefore, combining (22) and (23), we have as we wished to show:

$$\{x \in U : P(x)\}' = \{x \in U : \sim P(x)\}. \clubsuit$$

Theorem 21: Let U be the universal set, and $\varnothing$ the empty set. Then

1. $U' = \varnothing.$ (24)

2. $\varnothing' = U.$ (25)

Proof:

1. According to the definition of a complement, we have $U' = \{x \in U : x \notin U\}$. That is, U' is the set made up of those elements of U that are not elements of U. But since no such elements exist, we conclude that U' is empty. In other words, $U' = \varnothing$.

2. Similarly, $\varnothing' = \{x \in U : x \notin \varnothing\}$. Thus, noticing that the defining property $x \notin \varnothing$ is satisfied by all elements of U, we immediately see that $\varnothing' = U$. And the proof of the theorem is complete. ♣

5. 4. THE DIFFERENCE OF TWO SETS

We now introduce an operation that is very similar to complementation. This operation is known as the **difference operation.**

Definition 22: Let A and B be subsets of the universal set U. The **difference of** B **from** A, written as $A - B$, is the set of all those elements of U that belong to A but not to B.

Remark 23: Thus, the above definition can be written as:

$$A - B = \{x \in U : x \in A \text{ and } x \notin B\}. \qquad (26)$$

That is,

$$x \in A - B \Leftrightarrow x \in A \text{ and } x \notin B. \blacksquare \qquad (27)$$

Example 24:
Let $U = \{1, 2, 3, \ldots, 10\}$ be a universal set, $A = \{1, 2, 3, 4, 5\}$ and $B = \{4, 5, 6, 7, 8, 9\}$. Then describe $A - B$ and A'.

Solution:
Clearly

$$A - B = \{1, 2, 3\}$$

and

$$A' = \{6, 7, 8, 9, 10\}. \quad ♥$$

> **Theorem 25:** Let A and B be subsets of the universal set U. Then,
> $$A - B = A \cap B'.$$

Proof:

$$\begin{aligned} A - B &= \{x \in U : x \in A \textit{ and } x \notin B\} \\ &= \{x \in U : x \in A \textit{ and } \sim (x \in B)\} \\ &= \{x \in U : x \in A\} \cap \{x \in U : \sim (x \in B)\} \\ &= \{x \in U : x \in A\} \cap \{x \in U : x \in B\}' \\ &= A \cap B', \text{ as we wished to prove. } ♣ \end{aligned}$$

5. 5. THE CARTESIAN PRODUCTS OF TWO SETS

ORDERED PAIRS

We learned in Chapter 4 that the order in which the elements of a set are listed is of no significance in the definition of the set. Thus, for example, $\{2, 6\}$ and $\{6, 2\}$ both describe the same set. Yet, there are numerous situations in mathematics and everyday life where the specific order of items in a pair does matter. For instance, in coordinate geometry, we are able to locate every point M of the Cartesian plane by a pair of numbers, (x, y), where it is agreed that the first number x represents the horizontal coordinate, and the second number y the vertical coordinate. In this case, we know pretty well that the pair $(2, 6)$ designates a point different from that represented by $(6, 2)$, even though the two are made up of the same elements. Hence, in this context, order counts; and these pairs should therefore be called **ordered pairs**.

Definition 26: An **ordered pair of** elements x and y, denoted by (x, y), is a pair of elements where one of the elements, in this case x, is designated as first and the other, in this case y, is second. [Note the use of parenthesis as opposed to curly brackets.]

Remark 27: In other words, the order in which elements are listed in an ordered pair is crucial. Hence two ordered pairs (x, y) and (a, b) are equal, provided their first elements are equal, $x = a$, and their second elements are also equal $y = b$. ■

Example 28:
Let $M = \{1, 2, 3\}$ and $N = \{a, b\}$.

List all possible ordered pairs whose first elements belong to M, and second elements to N.

Solution:
There are two ordered pairs with first elements equal to 1:

$$(1, a), \quad (1, b)$$

There are two ordered pairs with first elements equal to 2:

$$(2, a), \quad (2, b)$$

There are two ordered pairs with first elements equal to 3:

$$(3, a), \quad (3, b)$$

Thus, we have all in all 3×2 possible ordered pairs. Therefore, the collection of all ordered pairs whose first elements come from M, and second elements from N is as follows:

$$\{ (1, a), \ (1, b), \ (2, a), \ (2, b), \ (3, a), \ (3, b) \}. \ ♥$$

CARTESIAN PRODUCTS

In the preceding example we found out that the collection of all ordered pairs whose first elements belong to set $M = \{1, 2, 3\}$ and second elements to set $N = \{a, b\}$ is the set:

$$\{ (1, a), (1, b), (2, a), (2, b), (3, a), (3, b) \}.$$

This set is called the **Cartesian product** of sets M and N, and is symbolized by $M \times N$.

Definition 28: Let A and B be any two sets. Then the **Cartesian product** of A with B, denoted by $A \times B$ and read "A cross B," is the set of all ordered pairs (x, y), whose first element x comes from A and second elements y from B.

Remark 29: Using set-builder notation, we may rewrite this definition compactly as:

$$A \times B = \{ (x, y) : \ x \in A, \ \text{and} \ y \in B \}. \ \blacksquare$$

Example 30:
Let $S = \{2, 3\}$ and $R = \{3\}$. Describe $S \times R$, $R \times S$, and $(R \times S) \cap (R \times S)$.

Solution:

$S \times R = \{ (2,3), (3,3) \}$,

$R \times S = \{ (3,2), (3,3) \}$,

$(R \times S) \cap (R \times S) = \{ (3,3) \}$. ♥

Theorem 31: Let A and B be any two sets. Then,

$$A \times B = \varnothing \quad \Leftrightarrow \quad A = \varnothing \ \textbf{or} \ B = \varnothing.$$

Proof:
To demonstrate the theorem, we need to prove two things:

The assertion

$$\text{If } A \times B = \varnothing, \text{ then } A = \varnothing \text{ or } B = \varnothing \tag{28}$$

and its converse

$$\text{If } A = \varnothing \text{ or } B = \varnothing, \text{ then } A \times B = \varnothing. \tag{29}$$

In both cases, we will use the method of **proof by contradiction**.

- **Proof of: If $A \times B = \varnothing$, then $A = \varnothing$ or $B = \varnothing$.**

Assume $A \times B = \varnothing$ is true, but $[A = \varnothing \text{ or } B = \varnothing]$ false. This means $A \times B = \varnothing$ is true and the denial $\sim [A = \varnothing \text{ or } B = \varnothing]$ is true. But this, in turn, is exactly the same thing as saying that the Cartesian product $A \times B$ is empty while the sets A and B are both non-empty.

Nowever, from the non-emptiness of the sets A and B, we know that there exists at least one item, say m, in A and at least one item, say n, in B. Thus, the Cartesian product $A \times B$ contains at least one element, namely the ordered pair (n, m); which implies that $A \times B$ is non-empty, which in turn clearly contradicts the assumption made earlier that $A \times B$ equals the empty set.

- **Proof of: If $A = \varnothing$ or $B = \varnothing$, then $A \times B = \varnothing$.**

Now, to show the second part of the theorem, let us suppose

$$[A = \varnothing \text{ or } B = \varnothing] \text{ is true},$$

while

$$A \times B = \varnothing \text{ is false}.$$

This means that at least one of the two sets, say for example A, is empty while the Cartesian product $A \times B$ is non-empty. But, from the non-emptiness of $A \times B$, we deduce that there is at least one ordered pair in $A \times B$. Let (s, r) be that ordered pair, then its first element s must belong to A. But, this implies A is not empty; a result which is in contradiction to the assumption made earlier that set A was empty. Notice that had we supposed that B, rather than A, was the empty set, the proof would have still been quite the same.

Thus this completes the proof of the theorem. ♣

EQUINUMEROUS SETS

To close this chapter, we will present two more important notions: the concept of a **one-to-one corresponce**, and that of **equinumerous sets**. These are pretty pervasive concepts in modern mathematics, and will be re-encountered in greater details later in this book.

Definition 32: Let A and B be two sets. A **one-to-one correspondence** between A and B is any pairing of the elements of A with those of B in such a way that each element of A corresponds to exactly one element of B, and conversely each element of B corresponds to exactly one element of A.

If a one-to-one correspondence can be set up between sets A and B, then we write $A \sim B$ and say that the two sets are **equivnumerous**.

Example 33:
Prove that every nonempty set A is equinumerous to itself.

Solution:
It suffices to point out the existence of at least one one-to-one correspondence between set A and itself. Clearly, one such correspondence is obtained by pairing each item x in set A with the item x itself. [Hence, this is a proof by construction of a one-to-one correspondence.] ♥

Remark 34: the foregoing example is a clear demonstration that equal sets are equinumerous sets. However, as is shown in the next example, two sets may be equinumerous without being equal. ■

Example 34:
Let $A = \{1, 2, 3, 4, 5\}$ and $V = \{a, e, i, o, u\}$
Prove that $A \sim V$.

Solution:
Here too, it suffices to set up a one-to-one correspondence between set A and set V. One such correspondence is given by the following pairing of the elements of A with those of V : $(1, a)$, $(2, e)$, $(3, i)$, $(4, o)$, and $(5, u)$. Indeed, notice that in this pairing every element of set A corresponds to one and only one element of set V, and vice versa. ♥

Example 35:
Let $\mathbb{N} = \{1, 2, 3, \ldots\}$, the set of all counting numbers, and $\mathbb{E} = \{2, 4, 6, \ldots\}$, the set of all even numbers. Show that $\mathbb{N} \sim \mathbb{E}$.

Solution:
To show that $\mathbb{N} \sim \mathbb{E}$, let us consider the paring $(n, 2n)$ of every element n of $\mathbb{N}$ with its double $2n$ in $\mathbb{E}$. Clearly, these pairings are a one-to-one correspondence between $\mathbb{N}$ and $\mathbb{E}$. For, indeed, each element of $\mathbb{N}$ corresponds to exactly one element of $\mathbb{E}$ (its double), and conversely each element of $\mathbb{E}$ corresponds to exactly one element of $\mathbb{N}$ (its half.) ♥

Remark 35 : Example 35 shows something quite counter-intuitive, which is that a set may be equinumerous to one of its proper subsets. As will be seen in Chapters 18 and 19, this property is true only in the case the set and its proper subset are so-called **infinite** sets. In other word, a **finite** set cannot be equinumerous with any of its proper subsets. ■

PROBLEMS

1. Let A, B and C be any sets. Then prove that

    1.1. $A \cap B = B \cap A$

    1.2. $A \cup B = B \cup A$

    1.3. $(A \cap B) \cap C = A \cap (B \cap C)$

    1.4. $(A \cup B) \cup C = A \cup (B \cup C)$

    1.5. $A \cup (B \cap C) = (A \cup B) \cap (A \cup C)$

    1.6. $A \cap (B \cup C) = (A \cap B) \cup (A \cap C)$

    1.7. $A \cup A = A$

    1.8. $A \cap A = A$.

2. Let A and B be any two subsets of a universal set U. Prove that

    2.1. $A \subset B \iff B' \subset A'$

    2.2. $(A \cap B)' = A' \cup B'$

    2.3. $(A \cup B)' = A' \cap B'$

    2.4. $A \cup U = U$

    2.5. $A \cap U = A$

    2.6. * $A \cup B = A \cup (B - A)$

    2.7. $A \subset B \iff A \cup B = B$

    2.8. $A \subset B \iff A \cap B = A$.

3. Rewrite these sets by using interval notations or the listing method

    3.1. $S = \{ x : \ x^2 + 8x + 7 = 0 \}$

    3.2. $S = \{ x : \ x^2 + 8x + 7 < 0 \}$

    3.3. $S = \{ x : \ x^2 + 8x + 7 > 0 \}$

3.4. $S = \{ x : \ x^2 + 8x + 7 \neq 0 \}$

4. Let A , B and C be any three subsets of a universal set U .

4.1. * Show that $A \subset B \Rightarrow A \cup C \subset B \cup C$

4.2. Show that $A \subset B \Rightarrow A \cap C \subset B \cap C$

5. Give a shorter and more elegant proof of Theorem 7. You may use the obvious facts that :

$$A = \{ x \in U : \ x \in A \},$$

$$B = \{ x \in U : \ x \in B \}$$

and

$$A \cap B = \{ x \in U : \ x \in A \ \ and \ \ x \in B \} .$$

6. Give a shorter and more elegant proof of Theorem 13. You may use the obvious facts that :

$$A = \{ x \in U : \ x \in A \},$$

$$B = \{ x \in U : \ x \in B \}$$

and

$$A \cup B = \{ x \in U : \ x \in A \ \ or \ \ x \in B \} .$$

"Mathematics as a science commenced when first someone, probably a Greek, proved propositions about "any" things or about "some" things, without specifications of definite particular things."
Alfred North Whitehead.

CHAPTER 6

SINGLE-VARIABLE SENTENTIAL[7] LOGIC

In the preceding chapter we introduced the concept of an open sentence and that of its solution set. Now, we will continue our study of these sentences.

We have already seen that one way to turn an open sentence into a proposition is to assign to its variable a specific value chosen from the universal set U. In this chapter, we will discuss another way of converting open sentences into propositions. This alternative way is commonly referred to as **quantification** of open sentences.

6.1. QUANTIFIED PROPOSITIONS

Consider the following open sentences:

$$(x-5)(x-2)=0, \quad x \in \mathbb{R}, \tag{1}$$

and

$$x^2+1>0, \quad x \in \mathbb{R}. \tag{2}$$

We known that the first one is true only **for some** values of $x \in \mathbb{R}$; namely, $x = 5$ and $x = 2$. As for the second sentence, it is clearly true

[7] **Sentential logic** is also called **predicate logic** by some authors

for all values of $x \in \mathbb{R}$. Therefore, unlike (1) and (2), the new sentences

$$\textbf{For some } x \in \mathbb{R}, \quad (x-5)(x-2) = 0 \tag{3}$$

and

$$\textbf{For all } x \in \mathbb{R}, \quad x^2 + 1 > 0 \tag{4}$$

are not open sentences but propositions [whose truth values are T.]

Note that in (3), we are saying that the truth set of our open sentence is nonempty; whereas in (4), we are claiming that the truth set is the entire universal set $\mathbb{R}$.

Definition 1: A phrases such as **"for some"** or **"for all,"** that may be used to claim that the truth set of an open sentence $P(x)$ is nonempty or the whole universal set U, is called a **quantifier**.

We shall refer to a proposition formed by combining an open sentence with a quantifier as a **quantified proposition**. Hence (3) and (4) are quantified propositions.

Remark 2: Note that instead of combining sentence (1) with the quantifier **"for some"** and sentence (2) with the quantifer **"for all,"** as we did earlier, we could have swapped these quantifiers, and append to sentence (1) the quantifier "for all" and to sentence (2) the quantifier "for some". Doing so would have given us the propositions:

$$\textbf{For all } x \in \mathbb{R}, \quad (x-5)(x-2) = 0 \tag{5}$$

and

$$\textbf{For some } x \in \mathbb{R}, \quad x^2 + 1 > 0. \tag{6}$$

Clearly, in this case we see that while proposition (6) is true, proposition (5) is false. ■

These examples show that, besides replacing the variable of an open sentence by a specific value, there is another way of turning that open sentence into a proposition. And as we just saw, this other way consists simply in quantifying the open sentence with either one of these phrases: "**for all**" or "**for some**".

6.2. UNIVERSAL PROPOSITIONS

Since the expression "**for all**", as used in the preceding discussions, turns open sentences into **general** or **universal** claims, it is called the **universal quantifier**. The universal quantifier is commonly represented by the symbol $\forall$, which is basically an inverted "A".

The symbol $\forall$ followed by a variable x [belonging to set U] can be read in one of several ways as shown below:

Symbolism	Possible Readings
$\forall x \in U$	For all $x \in U$
	For every $x \in U$
	For each $x \in U$
	For any $x \in U$

Example 3:

Translate these propositions into plain English, and tell their truth values.

1. $\forall x \in \mathbb{R}, \quad x^2 + 2x + 1 = (x+1)^2$.
2. $\forall n \in \mathbb{N}, \quad n - 1 > 0$.
3. $\forall p \in \mathbb{R}, \quad p - 1 > 0$.

Solution:

3.1. $\forall x \in \mathbb{R},\ x^2 + 2x + 1 = (x+1)^2$

may be read as

For every $x \in \mathbb{R},\ x^2 + 2x + 1 = (x+1)^2$,

which is a true quantified proposition since the open sentence involved in it is an identity.

3.2. $\forall n \in \mathbb{N},\ n - 1 > 0$

can be translated as

For all $n \in \mathbb{N},\ \ n - 1 > 0$,

which is false. Indeed, for example when $n = 1$, we have $n - 1 = 0$, a number that is clearly not greater than 0.

3.3. $\forall p \in \mathbb{R},\ p - 1 > 0$

may be read as

For any $p \in \mathbb{R},\ \ p - 1 > 0$,

which is a false proposition. Indeed, for $p = 0.12$ we have $p - 1 = -\ 0.88$, a number that is clearly not greater than 0. ♥

Definition 4: Let $R(x)$ be an open sentence about the variable $x \in U$. The quantified proposition $\forall x \in U, R(x)$ is called a **universally quantified proposition.**

We are now in the position to prove the following

Corollary 5: Let $R(x)$ be an open sentence about the variable x of the universal set U . Then:

$\left[\ \forall x \in U,\ R(x)\ \right]$ is true $\Leftrightarrow$ The truth set of $R(x)$ is the set U .

In other words, to say that a universally quantified proposition $\forall x \in U,\ R(x)$ is true is the same as claiming that the truth set of its open sentence $R(x)$ is the entire universal set U .

Proof:

It follows directly from of the meaning of the quantifier $\forall$. ♣

Corollary 6: A universally quantified proposition $\forall x \in U,\ R(x)$ is false, if and only if the truth set of its open sentence is a proper subset[8] of the universal set U .

Proof:

The proof follows readily from Corollary 5 by contraposition: $[p \Leftrightarrow q] \Leftrightarrow [\sim p \Leftrightarrow \sim q]$. See Problem 7 of Chapter 2 where we showed that $[p \Leftrightarrow q] \Leftrightarrow [\sim p \Leftrightarrow \sim q]$.

Example 7:

Which ones of these universally quantified propositions are true?

1. $\forall x \in \mathbb{R},\quad 8x + 4 + 3(x - 7) = 2(x + 2) + 9x - 21$
2. $\forall x \in \mathbb{R},\quad x^2 + 5x + 7 = 3(x + 2).$

Solution:

7.1. To find out whether this quantified proposition is true or false, we may first determine the true set of its open sentence.

We know that

$$8x + 4 + 3(x - 7) = 2(x + 2) + 9x - 21$$

is equivalent to

$$8x + 4 + 3x - 21 = 2x + 4 + 9x - 21\,. \tag{7}$$

And also, we know that

$$\begin{aligned}(7) &\Leftrightarrow 11x - 17 = 11x - 17\\ &\Leftrightarrow (11 - 11)x = 17 - 17\\ &\Leftrightarrow 0x = 0\,,\end{aligned}$$

[8] A set X is called a proper subset of a set Y ; if X is a subset of Y , but X does not contain all the element of Y.

which is satisfied by all $x \in \mathbb{R}$. Thus, the truth set in this case is the entire set $\mathbb{R}$. Therefore, by Corollary 5,

$$\forall x \in \mathbb{R}, \quad 8x + 4 + 3(x - 7) = 2(x + 2) + 9x - 21.$$

is a true proposition.

7.2. Similarly, to find out whether our second proposition is true or false, we will first determine the true set of the open sentence: $x^2 + 5x + 7 = 3(x + 2)$.

$$\begin{aligned} x^2 + 5x + 7 = 3(x+2) &\Leftrightarrow \quad x^2 + 5x + 7 = 3x + 6 \\ &\Leftrightarrow \quad x^2 + 2x + 1 = 0 \\ &\Leftrightarrow \quad (x+1)^2 = 0 \\ &\Leftrightarrow \quad x + 1 = 0 \\ &\Leftrightarrow \quad x = -1. \end{aligned}$$

Thus, the truth set in this case is $\{-1\}$ which is not the entire universal set $\mathbb{R}$, but only a proper subset of $\mathbb{R}$. Therefore, by Corollary 6,

$$\forall x \in \mathbb{R}, \ x^2 + 5x + 7 = 3(x+2)$$

is false. ♥

6.3. EXISTENTIAL PROPOSITIONS

Unlike "**for all**", the quantifier "**for some**" is not used for making universal claims. In fact, when "**for some**" is used, it is with the intention to assert the mere **existence** of some object that satisfies a given open sentence. For example, the proposition:

$$\text{For some } x \in \mathbb{R}, \quad 2x - 18 = 0 \tag{8}$$

should be understood as "There exists (an object) $x \in \mathbb{R}$, (such that) twice (that object) x minus 18 equals 0". For this reason, the phrase "**for some**" is rightly called the **existential** quantifier.

We shall represent the quantifier "**for some**" by the symbol $\exists$, which may be viewed as a turned-around "E". This symbol too may be read in one of several ways as shown below:

Symbolism	Possible Readings
$\exists x \in U$	For some $x \in U$
	There exists $x \in U$ (such that)
	There exists at least one $x \in U$ (such that)
	There is (**a fixed**) $x \in U$ (such that)
	There is at least one $x \in U$ (such that)

Example 8:
Translate these propositions into plain English, and tell their truth values.

1. $\exists x \in \mathbb{R}, \quad x^2 + 2x + 1 = (x+1)^2$.
2. $\exists n \in \mathbb{N}, \quad n - 1 > 20$.
3. $\exists p \in \mathbb{N}, \quad 2p = 1$.

Solution:
8.1. $\exists x \in \mathbb{R}, \quad x^2 + 2x + 1 = (x+1)^2$
may be read

$$\text{There is } x \in \mathbb{R}, \quad x^2 + 2x + 1 = (x+1)^2,$$

which is a true proposition. Indeed, since the open sentence $x^2 + 2x + 1 = (x+1)^2$ is true for any $x \in \mathbb{R}$, clearly there is at least a number $x \in \mathbb{R}$ for which it is true.

8.2. $\exists n \in \mathbb{N}, \quad n-1>0$

can be translated as

There is (an object) $n \in \mathbb{N}$ (such that) $n-1>0$,

which is definitely true. Indeed, for example, for $n=10$, we have $n-1=9$, which is greater than 0.

8.3. $\exists p \in \mathbb{N}, \quad 2p=1$

may be read

There exists $p \in \mathbb{N}, \quad 2p=1$,

which is clearly false. For, indeed, no matter which value we choose for p in the set of counting numbers $\mathbb{N}$, $2p$ is even. In other words, we cannot have $2p=1$. ♥

Definition 9: Let $R(x)$ be an open sentence about the variable $x \in U$. The proposition $\exists x \in U, R(x)$ is called an **existentially quantified proposition**.

Next, we prove the following claim.

Corollary 10: Let $R(x)$ be an open sentence about the variable $x \in U$. Then:

$[\exists x \in U, R(x)]$ is true $\Leftrightarrow$ The truth set of $R(x)$ is a nonempty subset of U.

In other words, to say that an existentially quantified proposition $\exists x \in U, R(x)$ is true is the same as to claim that the truth set of its open sentence $R(x)$ is a non-empty subset of set U.

Proof:

It is a direct consequence of the meaning of the quantifier $\exists$. ♣

Corollary 11: An existentially quantified proposition $\exists x \in U, R(x)$ is false, if and only if the truth set of its open sentence is the empty set $\varnothing$.

Proof:
The proof follows readily from Corollary 10, by contraposition: $[p \Leftrightarrow q] \Leftrightarrow [\sim p \Leftrightarrow \sim q]$. ♣

Example 12:
Which ones of these existentially quantified propositions are true?

1. $\exists x \in \mathbb{R}, \quad 8x + 4 + 3(x - 7) = 2(x + 2) + 9x - 21$.
2. $\exists x \in \mathbb{R}, \quad x^2 + 5x + 7 = 3(x + 2)$

Solution:
12.1. We know from the solution of Example 7.1 that the truth set of the open sentence

$$8x + 4 + 3(x - 7) = 2(x + 2) + 9x - 21$$

is the entire universal set $\mathbb{R}$, which is nonempty. Therefore, by virtue of Corollary 10,

$$\exists x \in \mathbb{R}, \ 8x + 4 + 3(x - 7) = 2(x + 2) + 9x - 21$$

is a true proposition.

12.2. Similarly, we know from Example 7.2 that the truth set of the open sentence

$$x^2 + 5x + 7 = 3(x + 2)$$

is the set $\{-1\}$, which is also nonempty. Hence,

$$\exists x \in \mathbb{R}, \ x^2 + 5x + 7 = 3(x + 2)$$

too is true. ♥

Remark 13: We just saw with Example 12 that the existentially quantified proposition

$$\exists x \in \mathbb{R}, \quad x^2 + 5x + 7 = 3(x + 2) \qquad (9)$$

is true, since its open sentence is satisfied by at least one element of the universal set $\mathbb{R}$. In fact, to be more precise, we should notice that the open sentence $x^2 + 5x + 7 = 3(x + 2)$ is satisfied **by one and only real number,** namely -1. To emphasize this uniqueness of the solution to the open sentence, we write

$$\exists!\, x \in \mathbb{R},\quad x^2 + 5x + 7 = 3(x + 2), \tag{10}$$

and read "There is a **unique** $x \in \mathbb{R}$ such that $x^2 + 5x + 7 = 3(x + 2)$." ■

6.4. DENIALS OF QUANTIFIED PROPOSITIONS

DENIALS OF UNIVERSAL PROPOSITIONS

Consider the universally quantified proposition:

$$\forall x \in U,\ R(x) \tag{11}$$

where $R(x)$ is an open sentence. In what follows, we will demonstrate that the denial of (11) is the existentially quantified proposition:

$$\exists x \in U,\ \sim R(x)\,. \tag{12}$$

We will do so by proving the following

Theorem 14: Let $R(x)$ be an open sentence about the variable x of the universal set U. Then

$$\sim [\, \forall x \in U,\ R(x)\,] \iff \exists x \in U,\ \sim R(x)\,. \tag{13}$$

Remark 15: In other words, to obtain the denial of a universally quantified proposition, we must simply replace its universal quantifier $\forall$ with the existential quantifier $\exists$, and deny its the open sentence. ■

Proof:

To show this theorem, it suffices to demonstrate that (11) and (12) have opposite truth values. In other words, we are going to show that if (11) is true, then (12) is false; and if (11) is false, then (12) is true.

- **Direct proof.** Assume the proposition (11) is true. Then by Corollary 5, we have:

$$\{\ x \in U :\ \ R(x)\ \} = U\ .$$

Thus,

$$\begin{aligned}\{\ x \in U :\ \sim R(x)\ \} &= \{\ x \in U :\ R(x)\ \}' \\ &= U' = \varnothing\ .\end{aligned}$$

In other words, the truth set of $\sim R(x)$ is the empty set $\varnothing$. Hence, by Corollary 6, the existentially quantified proposition (12) must be false.

- **Direct proof.** Now, suppose the universal proposition (11) is false. By Corollary 6, this means that the truth set of the open sentence $R(x)$ is not the entire universal set U, but some proper subset S of U.

Thus,

$$\begin{aligned}\{\ x \in U :\ \sim R(x)\ \} &= \{\ x \in U :\ R(x)\ \}' \\ &= S' \neq \varnothing\ .\end{aligned}$$

[Where the last inequality is owing to the fact that S is a proper subset of U.] In other words, the truth set of $\sim R(x)$ is non-empty. Hence, by Corollary 10, the existentially quantified proposition (12) is true. And this completes the proof. ♣

Example 16:
State the denials of these propositions:

1. $\forall n \in \mathbb{N},\quad 2n + 4 + 3(n - 7) + 6n = 2(n + 2) + 9n + 21.$
2. $\forall x \in \mathbb{R},\quad x^2 + 5x + 7 = 3(x + 2).$
3. $\forall x \in \mathbb{R},\quad x^2 + 1 > 0$
4. $\forall n \in \mathbb{N},\quad n^2 - 1 \geq 0$

Solution:

16.1. To find the denial, we replace the universal quantifier $\forall$ with the existential quantifier $\exists$, and deny the open sentence after the comma. We have:

$$\exists n \in \mathbb{N}, \quad \sim [2n + 4 + 3(n - 7) + 6n = 2(n + 2) + 9n + 21],$$

which is the same as:

$$\exists n \in \mathbb{N}, \quad 2n + 4 + 3(n - 7) + 6n \neq 2(n + 2) + 9n + 21.$$

16.2. In this case too, the denial is:

$$\exists x \in \mathbb{R}, \quad x^2 + 5x + 7 \neq 3(x + 2)$$

16.3. The denial is:

$$\exists x \in \mathbb{R}, \quad x^2 + 1 \leq 0.$$

16.4. The denial for this universally quantified proposition is:

$$\exists n \in \mathbb{N}, \quad n^2 - 1 < 0. \heartsuit$$

DENIALS OF EXISTENTIAL PROPOSITIONS

We now consider a typical existentially quantified proposition:

$$\exists x \in U, \quad R(x). \tag{14}$$

As in the preceding section, we want to show that the denial of the existentially quantified proposition (14) is the following universally quantified proposition:

$$\forall x \in U, \sim R(x) \tag{15}$$

Theorem 17: Let $R(x)$ be an open sentence about the variable x from the universal set U. Then

$$\sim [\exists x \in U, \ R(x)] \iff \forall x \in U, \sim R(x). \tag{16}$$

Remark 18: In other words, the denial of an existentially quantified proposition is formed by replacing its existential quantifier $\exists$ with the universal quantifier $\forall$, and by denying its open sentence. ■

Proof:

Here too, we just need to show that (14) and (15) have opposite truth values.

- **Direct proof.** Assume (14) is true. Then by Corollary 10 the truth set of the open sentence $R(x)$ is a nonempty subset S of the universal set U.

  Hence,

$$\{ x \in U : \sim R(x) \} = \{ x \in U : R(x) \}' = S' \neq U.$$

  In other words, the truth set of $\sim R(x)$ is not the entire universal set U. Hence, by Corollary 6, the universal proposition (15) is false.

- **Direct proof.** Now, assume that the existentially quantified proposition (14) is false. By Corollary 11, this means that the truth set of the open sentence $R(x)$ is the empty set $\varnothing$.

  Hence

$$\{ x \in U : \sim R(x) \} = \{ x \in U : R(x) \}' = \varnothing' = U$$

  In other words, the truth set of $\sim R(x)$ is the entire universal set U. Thus, by Corollary 5, the universal proposition (15) must be true. ♣

Example 19:

Write the denials of these propositions:

1. $\exists n \in \mathbb{N}, \quad 2(n-7) = 10.$
2. $\exists x \in \mathbb{R}, \quad x^2 + 2x + 1 = 0.$
3. $\exists x \in \mathbb{R}, \quad x^2 > 0.$
4. $\exists n \in \mathbb{N}, \quad n^2 - 1 \geq 0.$

Solution:

19.1. To find the denial, we replace the existential quantifier $\exists$ by universal quantifier $\forall$, and then we deny the open sentence. We have:

$$\forall n \in \mathbb{N}, \quad \sim[2(n-7)=10],$$

which is the same as

$$\forall n \in \mathbb{N}, \quad 2(n-7) \neq 10.$$

19.2. Here too, the denial is

$$\forall x \in \mathbb{R}, \quad x^2 - 2x + 1 \neq 0.$$

19.3. Here, the denial is simply

$$\forall x \in \mathbb{R}, \quad x^2 \leq 0.$$

19.4. The denial for this universally quantified proposition is:

$$\forall n \in \mathbb{N}, \quad n^2 - 1 < 0. ♥$$

PROBLEMS

1. Which ones of these quantified propositions are true and which ones are false?

   1.1. $\exists x \in \mathbb{Z},\ 2^x = 8.$

   1.2. $\forall x \in \mathbb{R},\ (x+2)^2 \neq x^2 + 2x.$

   1.3. $\forall x \in \mathbb{Z},\ 2^x = 8.$

   1.4. $\exists x \in \mathbb{R},\ (x+2)^2 = x^2 + 2x.$

2. Write the denial for each of the quantified propositions given in Problem 1 above, and tell which ones of these denials are true and which ones are false.

"All men by nature desire to know"
Aristotle (384-22BC)

CHAPTER 7

SENTENTIAL IMPLICATIONS AND EQUIVALENCES

We will continue in this chapter the study of open sentences begun in chapters 5 and 6.

7.1. SENTENTIAL IMPLICATIOBNS

We know from Problem 4 of chapter 2 that

$$p \rightarrow q \;\Leftrightarrow\; \sim p \vee q, \tag{1}$$

for any two propositions p and q.

This equivalence of the conditional $p \rightarrow q$ and the disjunction $\sim p \vee q$ is our motivation for following definition.

Definition 1: Let $R(x)$ and $Q(x)$ be open sentences about the variable x of the universal set U. Then by

$$R(x) \rightarrow Q(x)$$

we shall mean the open sentence

$$\sim R(x) \vee Q(x) .$$

We shall call the open sentence $R(x) \rightarrow Q(x)$ an **open conditional**.

Theorem 2: Let $R(x)$ and $Q(x)$ be open sentences about $x \in U$. Then the universally quantified proposition

$$\forall x \in U, \quad R(x) \rightarrow Q(x) \tag{2}$$

is true, if and only if the truth set of the open sentence $R(x)$ is a subset of the truth set of the open sentence $Q(x)$.

Proof:
From Corollary 5 of Chapter 6, we know that (2) is true, if and only if

$$\{x \in U : \ R(x) \rightarrow Q(x)\} = U .$$

Thus, to prove the present theorem, it suffices to show the following equivalence

$$\{x \in U : \ R(x) \rightarrow Q(x)\} = U \Leftrightarrow \{x \in U : \ R(x)\} \subset \{x \in U : Q(x)\} .$$

In other words, we must prove these two implications:

$$\{x \in U : R(x) \rightarrow Q(x)\} = U \Rightarrow \{x \in U : R(x)\} \subset \{x \in U : Q(x)\} \tag{3}$$

and

$$\{x \in U : R(x)\} \subset \{x \in U : Q(x)\} \Rightarrow \{x \in U : R(x) \rightarrow Q(x)\} = U . \tag{4}$$

- **Proof of (3):**
  Suppose $\{x \in U : \ R(x) \rightarrow Q(x)\} = U$.

  Then

$$\begin{aligned} U &= \{x \in U : \ R(x) \rightarrow Q(x)\} \\ &= \{x \in U : \ \sim R(x) \vee Q(x)\} \\ &= \{x \in U : \ \sim R(x)\} \cup \{x \in U : \ Q(x)\} \\ &= \{x \in U : \ R(x)\}' \cup \{x \in U : \ Q(x)\} . \\ &= Q \cup R' \end{aligned} \tag{5}$$

  where we have named R and Q the truth sets $\{x \in U : R(x)\}$ and $\{x \in U : Q(x)\}$, respectively.

Now, using **the method of contradiction**, it is easy to demonstrate that (5) means R is a subset of Q. Indeed, assume R is not a subset of Q. Then, there must be at least one element, say e, of set R which does not belong to set Q. Also, we know that $e \notin R'$; therefore $e \notin Q \cup R'$. This implies that U is not a subset of $Q \cup R'$ thereby contradicting (5).

- **Proof of (4):**

Conversely, now assume

$$\{x \in U : R(x)\} \subset \{x \in U : Q(x)\},$$

Hence, by using the result of Problem 2.1 of Chapter 5, we have

$$\{x \in U : Q(x)\}' \subset \{x \in U : R(x)\}' . \qquad (6)$$

But, from (6) we also deduce that

$$U = \{x \in U : Q(x)\}' \cup \{x \in U : Q(x)\}$$
$$\subset \{x \in U : R(x)\}' \cup \{x \in U : Q(x)\}.$$

where the latter relation is owing to (6) and the result of Problem 4.1 of Chapter 5.

Thus,

$$U \subset \{x \in U : \sim R(x)\} \cup \{x \in U : Q(x)\}$$
$$= \{x \in U : \sim R(x) \vee Q(x)\}$$
$$= \{x \in U : R(x) \to Q(x)\}.$$

Therefore,

$$U \subset \{x \in U : R(x) \to Q(x)\}. \qquad (7)$$

Furthermore, as U is our universe of discourse, we also have

$$\{x \in U : R(x) \to Q(x)\} \subset U. \qquad (8)$$

Thus, combining (7) and (8) we obtain

$$\{x \in U : R(x) \to Q(x)\} = U,$$

which means that the truth set of the open sentence $R(x) \to Q(x)$ is the entire universal set U .

Hence,

$$\forall x \in U, \quad R(x) \to Q(x),$$

is a true proposition.

And this completes the proof of our theorem. ♣

Example 3:
Which ones of these universally quantified propositions are true?

1. $\forall x \in \mathbb{R}, \quad x = 2 \to x^2 = 4.$
2. $\forall x \in \mathbb{R}, \quad x^2 = 4 \to x = 2.$

Solution:
3.1. This is a true proposition. Indeed, denoting by R the truth sets of $x = 2$ and by Q the truth set of $x^2 = 4$, we have $R = \{2\}$ and $Q = \{2, -2\}$; that is $R \subset Q$.

3.2. This is not a true proposition, since Q , the truth set of $x^2 = 4$, is not a subset of R, the truth set of $x = 2$. ♥

We now introduce the following notation.

Notation 4: Let $R(x)$ and $Q(x)$ be open sentences about the variable $x \in U$. Then, by

$$R(x) \Rightarrow Q(x)$$

we shall mean that

$$(\ \forall x \in U, R(x) \to Q(x)\ ) \text{ is true.}$$

The notation $R(x) \Rightarrow Q(x)$ is read " $R(x)$ **implies** $Q(x)$," and referred to as a **sentential implication**.

Remark 5: Thus, by virtue of Theorem 2, $R(x) \Rightarrow Q(x)$ if and only if the truth set of $R(x)$ is a subset of the truth set of $Q(x)$. Therefore, to establish a typical implication $R(x) \Rightarrow Q(x)$, we need only show that assuming the open sentence $R(x)$ to be true for an $x \in U$ leads to the open sentence $Q(x)$ also being true for that same $x \in U$. Thus, it is not necessary that one **first** determines the truth sets of $R(x)$ and $Q(x)$, and then verify that the truth set of $R(x)$ is contained in the truth set of $Q(x)$. ■

Example 6:
Using the above remark, prove: $x = 2 \Rightarrow x^2 = 4$.

Solution:
We do not have to find the truth sets of the open sentences $x = 2$ and $x^2 = 4$ as we did in the solution of Example 3. All we need is to let $x = 2$, and then show that this assumption leads, by force, to $x^2 = 4$, as we are now doing:

$$\begin{aligned} x = 2 &\Rightarrow x^2 = 2^2 \\ &\Rightarrow x^2 = 4. \end{aligned}$$ ♥

The following is a very important theorem; for it is frequently used in mathematics.

Theorem 7 [Law of Transitivity of Implications]: Let $R(x)$, $S(x)$ and $T(x)$ be open sentences about the variable $x \in U$. If

$$[R(x) \Rightarrow S(x) \text{ and } S(x) \Rightarrow T(x)]$$

then

$$R(x) \Rightarrow T(x).$$

Direct proof: Assume

$$R(x) \Rightarrow S(x) \qquad \text{and} \qquad S(x) \Rightarrow T(x).$$

This means

$$R \subset S \quad \text{and} \quad S \subset T$$

where R, S, and T are respectively the truth sets of $R(x)$, $S(x)$ and $T(x)$. Thus $S \subset T$, which automatically means that $R(x) \Rightarrow T(x)$. ♣

Remark 8: As an immediate consequence of Theorem 7, an implication $R(x) \Rightarrow Q(x)$ will be established, if one can exhibit a chain of simpler implications

$$R(x) \Rightarrow r_1(x) \Rightarrow r_2(x) \Rightarrow r_3(x) \Rightarrow \ldots \Rightarrow r_n(x) \Rightarrow Q(x),$$

that connects the open sentence $R(x)$ to the open sentence $Q(x)$.

Example 9: Show that $0 \le a \le b \Rightarrow a^2 \le b^2$.

Solution:
Let $0 \le a \le b$.

Therefore, multiplying all three sides of this inequality first by a, and then by b, we get:

$$\begin{aligned} 0 \le a \le b \quad &\Rightarrow \quad 0 \le a^2 \le ba \ \ \textit{and} \ \ 0 \le ab \le b^2 \\ &\Rightarrow \quad a^2 \le ba \quad \textit{and} \quad ab \le b^2 \\ &\Rightarrow \quad a^2 \le b^2. \end{aligned}$$

Notice that the principles used in establishing the above implications are well-known properties of real numbers: (i) The multiplication property of inequality[9], and (ii) The transitivity of the relation "$\le$ [10]". ♥

Corollary 10: If there is at least one value of $x \in U$ for which $R(x)$ is true and $Q(x)$ is false, then the implication $R(x) \Rightarrow Q(x)$ is false.

---

[9] If $x \le y$, and $z \ge 0$ then $xz \le yz$.

[10] If $x \le y$, and $y \le z$ then $x \le z$.

Proof:
First, recall that

$$\forall x \in U,\ R(x) \rightarrow Q(x) \qquad \text{and} \qquad \exists x \in U,\ \sim [R(x) \rightarrow Q(x)]$$

are denials of one another. Hence

$$R(x) \Rightarrow Q(x) \qquad \text{and} \qquad \exists x \in U,\ R(x) \wedge \sim Q(x)$$

are also denials of one another. Therefore if $\exists x \in U,\ R(x) \wedge \sim Q(x)$ is true, that is to say if there exists at least one value of $x \in U$ for which $R(x)$ is true while $Q(x)$ is at the same time false, then the implication $R(x) \Rightarrow Q(x)$ must be false . ♣

Example 11: Show that $x^2 = 4 \Rightarrow x = 2$ is false.

Solution:
By virtue of the foregoing corollary, $x^2 = 4 \Rightarrow x = 2$ is a false proposition. Indeed, there exists a value of x, namely $x = -2$, for which the open sentence $x^2 = 4$ is true, even as $x = 2$ is false. ♥

7.2. SENTENTIAL EQUIVALENCE

Definition 12: Let $R(x)$ and $Q(x)$ be open sentences about $x \in U$. Then by

$$R(x) \leftrightarrow Q(x)$$

we mean the open sentence

$$[R(x) \leftrightarrow Q(x)] \wedge [Q(x) \leftrightarrow R(x)].$$

We shall call the open sentence $R(x) \leftrightarrow Q(x)$ an **open biconditional**.

Next, we introduce the following

Notation 13: Assume $R(x)$ and $Q(x)$ are open sentences about the variable $x \in U$. Then, we shall write

$$R(x) \Leftrightarrow Q(x)$$

to mean

$$(\forall x \in U, R(x) \leftrightarrow Q(x)) \text{ is true.}$$

The notation $R(x) \Leftrightarrow Q(x)$ is read " $R(x)$ **is equivalent to** $Q(x)$," and referred to as a **sentential equivalence**."

We are now in the position to prove this claim

Theorem 14: Let $R(x)$ and $Q(x)$ be open sentences about $x \in U$. Then, we have the equivalence

$$R(x) \Leftrightarrow Q(x), \tag{9}$$

if and only if $R(x)$ and $Q(x)$ share the same truth set.

Direct proof:

$$\begin{aligned}
[R(x) \Leftrightarrow Q(x)] &\Leftrightarrow \big(\forall x \in U,\ R(x) \leftrightarrow Q(x) \big) \text{ is true} \\
&\Leftrightarrow \{x \in U : R(x) \leftrightarrow Q(x)\} = U \\
&\Leftrightarrow \{ x \in U : [R(x) \to Q(x)] \wedge [Q(x) \to R(x)] \} = U \\
&\Leftrightarrow \{ x \in U : R(x) \to Q(x) \} \\
&\qquad\qquad \cap \{ x \in U : Q(x) \to R(x) \} = U \\
&\Leftrightarrow \begin{cases} \{ x \in U : [Q(x) \to R(x)] \} = U \\ \text{and} \\ \{ x \in U : [R(x) \to Q(x)] \} = U \end{cases}
\end{aligned}$$

$$\Leftrightarrow \begin{cases} (\ \forall x \in U,\ Q(x) \to R(x)\ ) \text{ is true} \\ \text{and} \\ (\ \forall x \in U,\ R(x) \to Q(x)\ ) \text{ is true} \end{cases}$$

$$\Leftrightarrow \begin{cases} Q(x) \Rightarrow R(x) \\ \text{and} \\ R(x) \Rightarrow Q(x) \end{cases}$$

$$\Leftrightarrow \begin{cases} Q \subset R, \\ \text{and} \\ R \subset Q, \end{cases}$$

where Q, and R are respectively the truth sets of the open sentences $Q(x)$ and $R(x)$. Thus $R = Q$, which completes the proof. ♣

PROBLEMS

1. Which ones of these quantified propositions are true and which ones are false?

1.1. $\forall x \in \mathbb{R},\ x \in \mathbb{R} \to x^2 > 0$.

1.2. $\forall x \in \mathbb{R},\ x \in \mathbb{R} \to (x+1)^2 = x + 2x + 1$.

1.3. $\forall x \in \mathbb{R},\ 2x - 4 = 0 \to x^2 - 1 = 0$.

1.4. $\forall x \in \mathbb{R},\ x^2 - 5x + 6 = 0 \to x - 2 = 0$ *and* $x - 3 = 0$.

1.5. $\forall x \in \mathbb{R},\ x^2 - 5x + 6 = 0 \to x - 2 = 0$ *or* $x - 3 = 0$.

2. Which ones of these quantified propositions are true and which ones are false?

2.1. $\exists x \in \mathbb{R},\ x \in \mathbb{R} \leftrightarrow x^2 > 0$.

2.2. $\exists x \in \mathbb{R},\ x \in \mathbb{R} \leftrightarrow (x+1)^2 = x + 2x + 1$.

2.3. $\exists x \in \mathbb{R},\ 2x - 4 = 0 \leftrightarrow x^2 - 1 = 0$.

2.4. $\exists x \in \mathbb{R},\ x^2 - 5x + 6 = 0 \leftrightarrow x - 2 = 0$ *and* $x - 3 = 0$.

2.5. $\exists x \in \mathbb{R},\ x^2 - 5x + 6 = 0 \leftrightarrow x - 2 = 0$ *or* $x - 3 = 0$.

3. Which ones of these sentential inplications are true and which ones are false? [In each of these problems, we consider $\mathbb{R}$ as the universal set of the variable x.]

3.1. $x \in \mathbb{R} \Rightarrow x^2 > 0$.

3.2. $x \in \mathbb{R} \Rightarrow (x+1)^2 = x + 2x + 1$.

3.3. $2x - 4 = 0 \Rightarrow x^2 - 1 = 0$.

3.4. $x^2 - 5x + 6 = 0 \Rightarrow x - 2 = 0$ *and* $x - 3 = 0$.

3.5. $x^2 - 5x + 6 = 0 \Rightarrow x - 2 = 0$ *or* $x - 3 = 0$.

4. Which ones of these sentential equivalences are true and which ones are false? [In each of these problems, we consider $\mathbb{R}$ as the universal set of the variable x.]

4.1. $x \in \mathbb{R} \Leftrightarrow x^2 > 0$.

4.2. $x \in \mathbb{R} \Leftrightarrow (x+1)^2 = x + 2x + 1$.

4.3. $2x - 4 = 0 \Leftrightarrow x^2 - 1 = 0$.

4.4. $x^2 - 5x + 6 = 0 \Leftrightarrow x - 2 = 0$ *and* $x - 3 = 0$.

4.5. $x^2 - 5x + 6 = 0 \Leftrightarrow x - 2 = 0$ *or* $x - 3 = 0$.

"The race that does not value trained intelligence is doomed"
Alfred North Whitehead

CHAPTER 8

TWO-VARIABLE SENTENTIAL LOGIC

8.1. TWO-VARIABLE OPEN SENTENCES

As you might have noticed, so far we have discussed open sentences involving only one variable. That does not however mean that all open sentences are one-variable open sentences. In fact, in this chapter we will consider open sentences with two variables and study a few of their basic properties.

Consider the sentence:

$$xy > 10, \quad (1)$$

where x and y are counting numbers. As you know, in plain English, this sentence translates as follows:

"The product of counting numbers x and y is greater than 10,"

which is neither true or false until and unless one has chosen specific values for x and y. In fact, for $x = 3$ and $y = 2$, the sentence is false; whereas for $x = 5$ and $y = 7$, it is true. Thus, (1) is an example of an open sentence involving two variables.

Definition 1: An open sentence of two variables, say x and y, is a sentence about x and y, which becomes a proposition when both variables are replaced by specific values.

Remark 2: As with single-variable open sentences, it is important in the case of a two-variable open sentence that one specifies from the start of any discussion the replacement sets for both variables. We will denote by U_x and U_y the replacement sets for the variable x and y; respectively. In the remainder of this chapter, we shall often denote two-variable open sentences by $R(x,y)$ or $S(x,y)$. ■

Example 3:
Which ones of these are two-variable open sentences?

1. $x+2y=0$, $x\in]-10, 10[$, $y\in]2, \infty[$
2. $x+2y$, $x\in]2, 12[$, $y\in]5, 10[$
3. $(r-2)^2+(s-1)^2>16$, $r\in \mathbb{R}$, $s\in \mathbb{R}$
4. $7y^3=56$, $y\in \mathbb{R}$
5. $5=7$

Solution:
3.1. This is a two-variable open sentence in x and y.
3.2. This is not an open sentence, since to start with it is not even a sentence. Indeed, its plain English translation, "The sum of x and two y's", contains no conjugated verb. It is therefore just an [algebraic] expression in the variable x and y.
3.3. This is a two-variable open sentence in r and s.
4.4. This is a one-variable open sentence in y.
4.5. This is not an open sentence; it is a proposition; and clearly a false one. ♥

8.2. QUANTIFIED OPEN SENTENCES

We know from Chapter 6 that a single-variable open sentence in x may be converted into a proposition by simply appending to it a quantifier: $\forall x\in U$, or $\exists x\in U$. As we shall see in this section, the situation for two-variable open sentences is similar.

Consider the open sentence

$$x \geq y, \tag{2}$$

where, for the discussion, we will choose $U_x = U_y = \mathbb{N}$. Now, let us quantify (2) with respect to just one variable, say x, by using the universal quantifier. We get:

$$\forall x \in \mathbb{N}, \; x \geq y. \tag{3}$$

Now, it is not difficult to see that (3) which in plain English may be translated as

"Every counting number x is greater than or equal to
the counting number y."

is neither a true statement nor a false one. Indeed, the truth or falsity of (3) will depend on the value of the variable y: for $y = 1$ the statement is true; whereas for any other value of y, it is false. This shows that, in general, quantifying a two-variable open sentence with respect to one variable only (here x), does not yield a proposition. But instead, one gets a single-variable open sentence (here in the variable y).

Thus, we can make the following claim.

Theorem 4: In general, if $R(x, y)$ is a two-variable open sentence about $x \in U_x$ and $y \in U_y$, then quantifying $R(x, y)$ with respect to one variable alone does not produce a proposition. It produces instead one of the following single-variable open sentences:

$$\forall x \in U_x, \; R(x, y) \tag{4}$$

is an open sentence in the variable y only.

$$\exists x \in U_x, \; R(x, y) \tag{5}$$

is an open sentence in the variable y only.

$$\forall y \in U_y, \; R(x, y) \tag{6}$$

is an open sentence in the variable x only.

$$\exists y \in U_y, \; R(x, y) \tag{7}$$

is an open sentence in the variable x only.

In the singly-quantified sentence $\forall x \in U_x, R(x, y)$, we shall refer to the variable y as a **free variable**. In contrast, the variable x is named the **bound variable**. In other words, the free variable here is that variable with respect to which we have performed **no** quantification, whereas the bound variable is the variable with respect to which quantification has been performed. Also, we shall call **quantified open sentence** any sentence such as $\forall x \in U_x, R(x, y)$ in which quantification has not been performed on all variables.

8.3. QUANTIFIED PROPOSITIONS

Now, note that one way to turn the quantified sentence (3) into a proposition is to replace the free variable y by any specific value from its universal set U_y. For example, with y replaced by 1, we get a true proposition,

$$\forall x \in \mathbb{N}, \; x \geq 1, \tag{8}$$

whose truth is evident from its plain English translation:

"Every counting number x is greater than or equal to 1".

Theorem 5: In general, if $R(\alpha, \beta)$ is a two-variable open sentence involving the variables α and β, then any one of the singly-quantified sentences

$$\forall \alpha \in U_\alpha, \; R(\alpha, \beta), \qquad \exists \alpha \in U_\alpha, \; R(\alpha, \beta),$$

$$\forall \beta \in U_\beta, \; R(\alpha, \beta), \qquad \exists \beta \in U_\beta, \; R(\alpha, \beta)$$

is transformed into a proposition, if its **free variable** is assigned a specific value.

Example 6:

Turn these singly-quantified sentences into propositions. And tell the truth values of the resulting propositions.

1. $\forall x \in \mathbb{R}, \; x + y = 0, \quad y \in \mathbb{R}$

2. $\exists y \in \mathbb{R}, \quad x + y = 0 , \quad y \in \mathbb{R}$

Solution:

6.1. To turn this sentence into a proposition, we may set its free variable y to a specific value of our choice, say $y = 8$. In that case, our sentence becomes

$$\forall x \in \mathbb{R}, \quad x + 8 = 0 , \tag{9}$$

which translates as

"For every real number x, x plus 8 equals 0",

or simply as

"Every real number x plus 8 equals 0". (10)

But, (10) is clearly a false proposition; so too is (9).

6.2. Similarly, to turn this sentence into a proposition we may set its free variable x equal to a specific value of our choice, say $x = -2$. We then obtain

$$\exists y \in \mathbb{R}, \quad y - 2 = 0 , \tag{11}$$

which says,

"There exists a real number y which decreased by 2 equals 0".

But, that is clearly true, so is (11) then. ♥

Now, notice that instead of assigning a value to the free variable $y \in \mathbb{Z}$ in the quantified open sentence

$$\exists x \in \mathbb{Z} , \quad x > y , \tag{12}$$

one can as well manage to turn (12) into a proposition by preceding it with either $\exists y \in \mathbb{Z}$, or $\forall y \in \mathbb{Z}$.

Indeed, in the first case, we get

$$\exists y \in \mathbb{Z} , \exists x \in \mathbb{Z} , \quad x > y , \tag{13}$$

which, if translated as

"There is an integer y, and there is an integer x such that $x > y$"

clearly shows we have a true proposition. [Indeed, it suffices to choose $y = -2$ and $x = 2$ to convince ourselves of its truth.]

In the case where we choose to quantify (12) with $\forall y \in \mathbb{Z}$ instead, we must have

$$\forall y \in \mathbb{Z}, \ \exists x \in \mathbb{Z}, \quad x > y, \tag{14}$$

which means

"For any integer y, there is an integer x, such that $x > y$".

Note that this latter proposition too is true. Indeed, for any given integer y, it suffices to pick $x = y + 1$ in order to guarantee that $x > y$.

It is important to understand that while y in (14) may assume any integer value, once that value is chosen, the corresponding value for x cannot be arbitrarily picked. Indeed, as seen above, a right value for the variable x, say $y + 1$, is clearly dependent on that of y.

Remark 7: Suppose we are given a true doubly-quantified proposition of the form

$$\forall \alpha \in U_\alpha, \ \exists \beta \in U_\beta, \ R(\alpha, \beta),$$

where the first quantifier is a universal quantifier, and the second an existential quantifier. Then, in general, for every α, the corresponding value of β may depend on the value of α. To stress this dependence of β on α, we will sometimes write

$$\forall \alpha \in U_\alpha, \ \exists \beta(\alpha) \in U_\alpha, \ R(\alpha, \beta)$$

and read "For every α in U_α, there is a β, **dependant on** α, in U_β such that $R(\alpha, \beta)$ is true." ■

Example 8:

Translate these quantified propositions into plain English, and tell their truth values.

1. $\forall x \in \mathbb{R}, \exists y \in \mathbb{R}, \quad x + y = 0.$
2. $\exists y \in \mathbb{R}, \forall x \in \mathbb{R}, \quad x + y = 0.$

Solution:

8.1. This proposition may be translated as

"For every real number x, there is a real number y, such that the sum $x+y$ equals 0".

In other words, it says, given any number x whatsoever, one can find a number y for which the sum $x+y$ of both numbers is 0. Clearly, this is a true proposition. Indeed, for every number x, it suffices to choose the number y equal to $-x$. [Note that the value of y depends on that of x.]

8.2. This proposition can be translated as

"There is a (**fixed**) real number y, such that for every real number x, the sum $x+y=0$".

or simply as

"There is a number y whose sum with any number x equals 0".

It is therefore clearly false. ♥

We generalize the foregoing result into a theorem.

Theorem 9: Let $R(x,y)$ be an open sentence about the variables $x \in U_x$ and $y \in U_y$. Then quantifying $R(x,y)$, with respect to both variables, results in a proposition. We shall call such a proposition a **quantified proposition**.

8.4. SWAPPING QUANTIFIERS

In Example 6 of the preceding section we considered the quantified propositions:

$$\forall x \in \mathbb{R},\ \exists y \in \mathbb{R},\quad x+y=0. \tag{15}$$

$$\exists y \in \mathbb{R},\ \forall x \in \mathbb{R},\quad x+y=0. \tag{16}$$

Clearly, the two propositions differ only in that either of them may be obtained by swapping the positions of $\forall x \in \mathbb{R}$ and $\exists y \in \mathbb{R}$ in the other proposition. We also saw in that same Example 6 that while (15) is a true proposition, (16) is a false one.

Thus, in general, by swapping the position of a universal quantifier, $\forall x \in U_x$, with that of an existential quantifier, $\exists y \in U_y$, we end up with propositions whose truth values may be opposite.

Hence the following

Theorem 10: Let $R(x, y)$ be an open sentence about the variables x and y . Then

1. The truth of

$$\forall x \in U_x,\ \exists y \in U_y,\ R(x, y)$$

does not guarantee the truth of

$$\exists y \in U_y,\ \forall x \in U_x,\ R(x, y).$$

2. But the truth of

$$\exists x \in U_x,\ \forall y \in U_y,\ R(x, y)$$

implies that of

$$\forall y \in U_y,\ \exists x \in U_x,\ R(x, y).$$

Proof

10.1. The first part of the theorem has already been proved. Indeed, in Example 6 we saw that if the universal quantifier is swapped with the existential quantifier, then from the true proposition (15) we get the false proposition (16).

10.2. Now, suppose

$$\exists x \in U_x,\ \forall y \in U_y,\ R(x, y)$$

is a true proposition.

Then, this means there is at least **one fixed** $x_0 \in U_x$ such that

$$\forall y \in U_y, \quad R(x_0, y)$$

is true. But, the fact that this **fixed** $x_0 \in U_x$ works with every $y \in U_y$ to make the above proposition true means: for every $y \in U_y$, there is at least one $x \in U_x$ [in this case the **fixed** element x_0] such that $R(x, y)$ is true. Hence,

$$\forall y \in U_y, \exists x \in U_x, \ R(x, y)$$

is true as we wished to show. ♣

Example 11:
Translate the following quantified propositions into plain English, and tell the truth value of each of them.

1. $\forall x \in \mathbb{R}, \exists y \in \mathbb{R}, \quad x^2 + y = 0.$
2. $\exists x \in \mathbb{R}, \ \forall y \in \mathbb{R}, \quad x^2 + y = 0.$

Solution:
11.1. This proposition may be translated as:

"For any real number x, there is a real number y, such that the square, x^2, of the first number plus the second number y equals 0".

This proposition is true. Indeed, for each real number x, it suffices to choose for y the number $-x^2$; and the proposition automatically will be true.

11.2. We translate the other proposition as.

"There exists a **fixed** real number x, such that for all real number y, $x^2 + y = 0$."

Note the addition of the term "**fixed**." It is a convenient thing to include the term "**fixed**" whenever, in a doubly-quantified proposition, the existential quantifier appears before the universal quantifier. Indeed, it [serves as a tool that] helps us grasp the meaning of the proposition more easily.

Now, we can easily see that our proposition is false; for there is no such fixed number whose square added to any number whatsoever yields 0. ♥

8.5. DENYING DOUBLY-QUANTIFIED PROPOSITIONS

Let us consider the doubly-quantified proposition:

$$\forall x \in U_x, \ \ \exists y \in U_y, \ \ R(x, y). \tag{17}$$

We want to write the denial of (17). To do so, it is convenient to first rewrite (17) as follows:

$$\forall x \in U_x, \ \left[\ \exists y \in U_y, \ \ R(x, y) \ \right]. \tag{18}$$

Notice that the expression within the square brackets is an open sentence in the free variable x. Thus, applying both Theorems 14 and 17 of Chapter 6, we get:

$$\sim \left(\ \forall x \in U_x , \ \left[\ \exists y \in U_y , \ \ R(x, y) \right] \right)$$

is equivalent to

$$\exists x \in U_x , \ \ \sim \left[\ \exists y \in U_y , \ \ R(x, y) \right]$$

which is equivalent to

$$\exists x \in U_x , \ \ \left[\ \forall y \in U_y , \ \sim R(x, y) \right]$$

which in turn is

$$\exists x \in U_x , \ \ \forall y \in U_y , \ \sim R(x, y) .$$

Therefore, we see that the denial of $\forall x \in U_x, \exists y \in U_y, R(x, y)$ is the proposition:

$$\exists x \in U_x , \ \ \forall y \in U_y , \ \sim R(x, y) .$$

Hence, we have proved the following

Theorem 12: If $R(x, y)$ is an open sentence in the variables $x \in U_x$, and $y \in U_x$, then we have:

1. The denial of $\forall x \in U_x, \exists y \in U_y, R(x, y)$ is

$$\exists x \in U_x, \quad \forall y \in U_y, \quad \sim R(x, y).$$

2. The denial of $\exists x \in U_x, \forall y \in U_y, R(x, y)$ is

$$\forall x \in U_x, \quad \exists y \in U_y, \quad \sim R(x, y).$$

Example 13:
In the Calculus, one learns that a function f is continuous at a point $x = a$, if:

$$\forall \varepsilon > 0, \quad \exists \delta > 0,$$
$$\forall x \in \mathbb{R}, \quad x \in]a - \delta, \ a + \delta[\ \rightarrow \ f(x) \in]f(a) - \varepsilon, \ f(a) + \varepsilon[\ .$$

Give in symbolic form the denial of the proposition: f **is continuous at point** $x = a$.

Solution:
Using the preceding theorem, we get

$$\exists \varepsilon > 0, \quad \forall \delta > 0,$$
$$\sim \left[\forall x \in \mathbb{R}, \quad x \in]a - \delta, \ a + \delta[\ \rightarrow \ f(x) \in]f(a) - \varepsilon, \ f(a) + \varepsilon[\right]$$

which is the same as

$$\exists \varepsilon > 0, \quad \forall \delta > 0,$$
$$\exists x \in \mathbb{R}, \ \sim \left(x \in]a - \delta, \ a + \delta[\ \rightarrow \ f(x) \in]f(a) - \varepsilon, \ f(a) + \varepsilon[\right)$$

which in turn is equivalent to

$$\exists \varepsilon > 0,\ \forall \delta > 0,$$
$$\exists x \in \mathbb{R},\ \sim\left(\sim x \in]a-\delta,\ a+\delta[\ \text{ or }\ f(x) \in]f(a)-\varepsilon,\ f(a)+\varepsilon[\right).$$

Thus, the symbolic form of the negation of the proposition " f is continuous at $x = a$" is:

$$\exists \varepsilon > 0,\ \forall \delta > 0,$$
$$\exists x \in \mathbb{R},\ x \in]a-\delta,\ a+\delta[\ \text{ and }\ f(x) \notin]f(a)-\varepsilon,\ f(a)+\varepsilon[\ .$$ ♥

8.6. TRUTH SETS OF TWO-VARIABLE OPEN SENTENCES

In this section, we will define the truth set for a two-variable open sentence

Definition 14: Let $R(x, y)$ be an open sentence about the variables $x \in U_x$ and $y \in U_x$. Then by the truth set of $R(x, y)$, we mean the set of all ordered pairs (α, β) for which $R(\alpha, \beta)$ is a true proposition. In other words, the truth set of the open sentence $R(x, y)$ is :

$$\left\{ (\alpha, \beta) \in U_x \times U_y : \quad R(\alpha, \beta) \text{ is true} \right\}$$

Example 15:

Find the truth set of the open sentences

1. $x + y = 0$, $\quad x \in \mathbb{R},\ y \in \mathbb{R}$.
2. $x^2 + y^3 = 0$, $\quad x \in \mathbb{R},\ y \in \mathbb{R}$.

Solution:

15.1. Let S be the truth set of this sentence. Then:

$$S = \left\{ (x, y) \in \mathbb{R}\times\mathbb{R} : \quad x + y = 0 \right\}$$
$$= \left\{ (x, y) \in \mathbb{R}\times\mathbb{R} : \quad x = -y \right\}$$

$$= \{\ (-y,\ y):\ \ y \in \mathbb{R}\ \}.$$

15.2. Let S be the truth set of this sentence. Then:

$$S = \left\{ (x, y) \in \mathbb{R} \times \mathbb{R} : \quad x^2 + y^3 = 0 \right\}$$

$$= \left\{ (x, y) \in \mathbb{R} \times \mathbb{R} : \quad -x^2 = y^3 \right\}$$

$$= \left\{ (x, y) \in \mathbb{R} \times \mathbb{R} : \quad -\sqrt[3]{x^2} = y \right\}$$

$$= \left\{ (x, -\sqrt[3]{x^2}) : \quad x \in \mathbb{R} \right\}. \quad ♥$$

PROBLEMS

1. Let A and B be sets of real numbers, and let $f : \mathbb{R} \to \mathbb{R}$ be a function. Then rewrite the following propositions using the quantifier symbols $\exists$ and $\forall$:

    1.1. For every $x \in A$, there is a $y \in A$ such that $y > x$.

    1.2. There is an $x \in A$ such that, for every $y \in A$, $y > x$.

    1.3. For every $y \in \mathbb{R}$, there is an $x \in \mathbb{R}$ such that $f(x) = y$.

    1.4. For every $x \in \mathbb{R}$, and for every $y \in \mathbb{R}$, if $f(x) = f(y)$ then $x = y$.

    1.5. For every positive real number ε, there is a positive real number δ such that, for all $x \in \mathbb{R}$, and for all $y \in \mathbb{R}$, if $|x - y| < \delta$ then $|f(x) - f(y)| < \delta$.

2. After you have finished rewriting the propositions in Problem 1 above using the symbols of quantifiers, deny each of them.

3. After you have finished with Problem 2, rewrite every one of those denials in plain English without the symbols $\exists$ and $\forall$.

4. Write the denials of these doubly quantified propositions

4.1. $\forall x \in U_x, \ \forall y \in U_y, \ R(x, y)$.

4.2. $\exists x \in U_x, \ \exists y \in U_y, \ R(x, y)$.

"Mathematics, in its widest significance, is the development of all types of formal, necessary, deductive reasoning."
Alfred North Whitehead (1861-1947).

"It is vain to do with more what can be done with less."
William of Ockham (1285-1349).

CHA PTER 9

REAL NUMBERS
DEFINITION AND PROPERTIES

In this chapter we will introduce the **system of real numbers,** and study some of its properties. We will take as our starting point the little knowledge you already possess about this number system.

You have certainly heard about the real numbers. For, indeed, in your previous studies of elementary algebra, you did learn how to picture real numbers as points of a line. You discovered that there is a simple way to match the real numbers with the points on a straight line. This matching was done in such a manner as to have every real number paired with one and only one point, and every point of the line associated with one and only one real number. In other words, you accepted as a fact the existence of a **one-to-one correspondence** between the real numbers and the points of any given line.

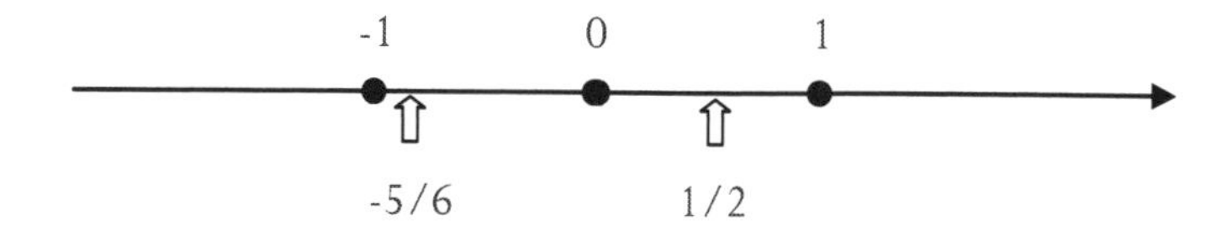

You called such a line [whose points have been paired with real numbers in a one-to-one correspondence] a **number-line**. Thus, you got into the habit of using the terms **point** and **real number** interchangeably. Also, you may recall that, by the very fashion in which you constructed your number-line, a real number x is less than another real number y , if the point associated to x lies to the left of the point associated to y .

Another important thing you also certainly learned is that the point associated with 0, and called the **origin** of the number-line, can be thought of as a **reference point** for two reasons:

- Any real number x is the measure of the line segment from the origin to the point associated with x .
- The number 0 splits the real numbers into two sets: The numbers that lie strictly to the left of the 0 form the **negative real numbers**; whereas the numbers strictly to the right of 0 make up the **positive real numbers**. As for 0, it is therefore neither positive nor negative; it is **neutral** so to speak.

Thus, we can assume that you have at least some ideas about the real numbers. Our task in this chapter is therefore not to present the real numbers merely as a set, but as a system. In mathematics, a **system** is more than just a set. The term system is generally used to designate a set on which we have defined (1) one or more operations, (2) one or more relations expressing some kind of comparison among elements of the set, and (3) a list of axioms that govern the behavior of the elements under these operations and relations.

9.1. THE SYSTEM OF REAL NUMBERS

To define the system of real numbers, we will take for granted the basic properties of real numbers concerning the operations of addition and multiplication, and those concerning the relations *is equal to* "=" and *is less than* "<." Instead of proving these basic properties, we shall consider them as **defining axioms**. Hence, ours is an **axiomatic approach** to introducing the system of real numbers. To be sure, there are others approaches[11] mainly by construction, but the advantage here is that the axiomatic approach permits a quicker, easier and most elegant access to main results concerning the real numbers.

The real number system, denoted by $(\mathbb{R}, =, +, \bullet, <)$, is the set $\mathbb{R}$ of real numbers equipped with the operations of addition " $+$ " and multiplication " $\bullet$ ", and the relations of equality " $=$" and order "$<$". Thus, as part of the definition of $(\mathbb{R}, =, +, \bullet, <)$ we will require that the following twenty axioms hold true.

EQUALITY AXIOMS [Axiom1-Axiom4]

The first axioms to be adopted with the system of real numbers have to do with the **equality relation**. In mathematics, a statement of equality is simply an assertion that two symbols, that are being equated, are names for exactly the same mathematical object.

Axiom1. Reflexivity for equality: For any number $a \in \mathbb{R}$,

$$a = a.$$

Axiom2. Symmetry for equality: For any a and b in $\mathbb{R}$,

$$\text{if } a = b, \text{ then } b = a.$$

Axiom3. Transitivity for equality: For any a, b and c in $\mathbb{R}$,

$$\text{if } a = b \text{ and } b = c, \text{ then } a = c.$$

Axiom4. Substitution: For any a and b in $\mathbb{R}$,

if $a = b$, then a may be replaced by b (or b by a) in any mathematical proposition without changing the truth value of that proposition.

---

[11] Dedekind cuts, and decimal expansions

ADDITION AXIOMS [Axiom5-Axiom9]

Axiom5. Closure for addition: The set $\mathbb{R}$ is closed with respect to addition. In other words,

if a and b are real numbers, then so is their sum $a+b$.

Axiom6. Commutativity for addition: For every a and b in $\mathbb{R}$,

$$a+b=b+a.$$

Axiom7. Associativity for addition: For every a, b and c in $\mathbb{R}$,

$$(a+b)+c=a+(b+c).$$

Axiom8. Existence of an identity for addition: There is in $\mathbb{R}$ (at least) one **fixed** element, denoted by 0, such that for every $x \in \mathbb{R}$

$$x+0=0+x=x.$$

The identity element for addition is sometimes called the **zero element** of $\mathbb{R}$.

Axiom9. Existence of an additive inverse of every element: For each element $x \in \mathbb{R}$, there exists an element in $\mathbb{R}$ called the additive inverse [or opposite] of x, and denoted by $-x$, such that

$$x+(-x)=(-x)+x=0.$$

MULTIPLICATION AXIOMS [Axiom10-Axiom14]

Axiom10. Closure for multiplication: The set $\mathbb{R}$ is closed with respect to multiplication. In others words,

if a and b are real numbers, then so is their product $a \bullet b$.

Axiom11. Commutativity for multiplication: For every a, and b in $\mathbb{R}$,

$$a \bullet b = b \bullet a.$$

Axiom12. Associativity for multiplication: For every a, b, and c in $\mathbb{R}$,

$$(a \bullet b) \bullet c = a \bullet (b \bullet c).$$

Axiom13. Existence of an identity for multiplication: There is in $\mathbb{R}$ (at least) one (fixed) element, denoted by 1, such that for every $x \in \mathbb{R}$

$$x \bullet 1 = 1 \bullet x = x.$$

The identity element for multiplication is sometimes called the **unit element** of $\mathbb{R}$.

Axiom14. Existence of a multiplicative inverse for every nonzero element: For each real number $x \neq 0$, there exists an element in $\mathbb{R}$ called the multiplicative inverse [or reciprocal] of x, and denoted by x^{-1}, such that

$$x \bullet x^{-1} = x^{-1} \bullet x = 1.$$

THE DISTRIBUTIVE AXIOM [Axiom15]

Axiom15. Distributivity of multiplication over addition: For every a, b, and c in $\mathbb{R}$,

$$a \bullet (b + c) = a \bullet b + a \bullet c,$$

and

$$(b + c) \bullet a = b \bullet a + c \bullet a.$$

ORDER AXIOMS [Axiom16-Axiom19]

Axiom16. Trichotomy: For any two numbers a and b in $\mathbb{R}$, one and only one of these three propositions is true:

$$a < b, \qquad a = b, \qquad b < a.$$

Axiom17. Transitivity: For every a, b, and c in $\mathbb{R}$,

$$\text{if } a < b, \quad \text{and} \quad b < c, \quad \text{then} \quad a < c.$$

Axiom18. Addition property of order: For every a, b and c in $\mathbb{R}$,

$$\text{if } a < b, \quad \text{then} \quad a + c < b + c.$$

Axiom19. Multiplication property of order: For every a, b and c in $\mathbb{R}$,

$$\text{if } a < b, \quad \text{and} \quad 0 < c \quad \text{then} \quad a \bullet c < b \bullet c.$$

THE CONTINUITY AXIOM [Axiom20]

The presentation of this axiom requires some new ideas, which we have yet to discuss. In other words, there are some preparatory works to be done first before we can smoothly present the axiom of continuity, and properly study its consequences. We shall, therefore, postpone the statement and study of this axiom till in Chapter 11.

9.2. SOME CONSEQUENCES OF THE ALGEBRAIC AXIOMS[12]

In this section, we shall establish some theorems of the real number system. As will be seen, these properties are sole consequences of the first fifteen axioms [of addition and multiplication]. They are therefore algebraic properties of real numbers.

Remark 21: In the remainder of this text, wherever it is more convenient, we shall denote multiplication, $a \bullet b$, simply by juxtaposition, ab. ■

[12] Algebraic axioms are the ones that involve the operations of addition and multiplication. They are Axiom1 through Axiom 15

Theorem 22: For every $a \in \mathbb{R}$, $0a = a0 = 0$.

Proof:

$$\begin{aligned}
0a &= a0 && \text{by Axiom 11.} \\
&= a0 + 0, && \text{by Axiom 8.} \\
&= a0 + [a + (-a)], && \text{by Axiom 9.} \\
&= [a0 + a] + (-a), && \text{by Axiom 7.} \\
&= [a0 + a1] + (-a), && \text{by Axiom 13.} \\
&= a[0 + 1] + (-a), && \text{by Axiom 15.} \\
&= a1 + (-a), && \text{by Axiom 8.} \\
&= a + (-a), && \text{by Axiom 13.} \\
&= 0 && \text{by Axiom 9 .} \clubsuit
\end{aligned}$$

Exercise 23: Explain why zero, 0, has no reciprocal in $\mathbb{R}$.

Solution

We will argue **by contradiction**. Suppose 0 had a reciprocal, say r. Then, by Axiom14, we should have $0r = 1$. But, Theorem 22 also implies that $0r = 0$. Thus, by Axiom 2, we have $1 = 0r = 0$. Therefore, by Axiom 3, we also have $1 = 0$ which is clearly a contradiction. ♥

Theorem 24: For every $a \in \mathbb{R}$, $a(-1) = (-1)a = -a$.

Proof:

$$\begin{aligned}
a(-1) &= (-1)a && \text{by Axiom 11.} \\
&= (-1)a + 0, && \text{by Axiom 8.} \\
&= (-1)a + [a + (-a)], && \text{by Axiom 9.}
\end{aligned}$$

$= [(-1)a + a] + (-a),$ by Axiom 7.

$= [(-1)a + 1a] + (-a),$ by Axiom 13.

$= [(-1) + 1]a + (-a),$ by Axiom 15.

$= 0a + (-a),$ by Axiom 9.

$= 0 + (-a),$ by Theorem 22.

$= -a,$ by Axiom 8. ♣

Exercise 25: Show that $-0 = 0$.

Solution

$-0 = (-1)0$ by Theorem 24

$= 0$, by Theorem 22. ♣

Theorem 26: For every a and b in $\mathbb{R}$, $(-a)b = a(-b) = -(ab)$

Proof:

$(-a)b$ * $= [(-1)a]b,$ by Theorem 24.

$= [a(-1)]b,$ by Axiom 11.

$= a[(-1)b],$ by Axiom 12.

$= a(-b),$ * by Theorem 24.

$= a[(-1)b],$ by Theorem 24.

$= a[b(-1)]$ by Axiom 11.

$= [ab](-1)$ by Axiom 12.

$= (-1)[ab]$ by Axiom 11.

$= -[ab]$ by Theorem 24

$= -ab$ * by order of operations . ♣

Theorem 27: For every a and b in $\mathbb{R}$, $(-a)(-b) = ab.$

Proof:

$$
\begin{aligned}
(-a)(-b) &= (-a)(-b)+0, && \text{by Axiom 8.}\\
&= (-a)(-b)+[\,ab+[-(ab)]\,], && \text{by Axiom 9.}\\
&= (-a)(-b)+[\,[-(ab)]+ab\,], && \text{by Axiom 6.}\\
&= (-a)(-b)+[\,[(-a)b]+ab\,], && \text{by Theorem 26.}\\
&= [\,(-a)(-b)+(-a)b\,]+ab, && \text{by Axiom 7.}\\
&= [\,(-a)[(-b)+b]\,]+ab, && \text{by Axiom 15.}\\
&= [\,(-a)0\,]+ab, && \text{by Axiom 9.}\\
&= 0+ab, && \text{by Theorem 22}\\
&= a\,b && \text{by Axiom 8.} \clubsuit
\end{aligned}
$$

Theorem 28: For every $a \in \mathbb{R}$, $-(-a) = a.$

Proof:

$$
\begin{aligned}
-(-a) &= -(-a)+0, && \text{by Axiom 8.}\\
&= -(-a)+[\,a+(-a)\,], && \text{by Axiom 9.}\\
&= (-1)(-a)+[\,a+(-a)\,], && \text{by Theorem 24.}\\
&= (-1)(-a)+[\,a+(-1)a\,], && \text{by Theorem 24.}\\
&= (-1)(-a)+[\,(-1)a+a\,], && \text{by Axiom 6.}\\
&= [\,(-1)(-a)+(-1)a\,]+a, && \text{by Axiom 7.}\\
&= (-1)[\,(-a)+a\,]+a, && \text{by Axiom 15.}\\
&= (-1)0+a && \text{by Axiom 9}\\
&= 0+a && \text{by Theorem 22}\\
&= a && \text{by Axiom 8.} \clubsuit
\end{aligned}
$$

Theorem 29: For every real number $a \neq 0$, $(a^{-1})^{-1} = a$

Proof:

$$
\begin{aligned}
(a^{-1})^{-1} &= 1(a^{-1})^{-1} && \text{by Axiom 13.} \\
&= \left[aa^{-1}\right](a^{-1})^{-1} && \text{by Axiom 14.} \\
&= a\left[a^{-1}(a^{-1})^{-1}\right] && \text{by Axiom 12.} \\
&= a1, && \text{by Axiom 14.} \\
&= a && \text{by Axiom 13.} \clubsuit
\end{aligned}
$$

Theorem 30: For every a and b in $\mathbb{R}$, $(ab)^{-1} = a^{-1}b^{-1}$.

Proof:

$$
\begin{aligned}
(ab)^{-1} &= \left[(ab)^{-1}1\right]1 && \text{by Axiom 13.} \\
&= \left[(ab)^{-1}(aa^{-1})\right](bb^{-1}), && \text{by Axiom 14.} \\
&= \left[\left[(ab)^{-1}a\right]a^{-1}\right](bb^{-1}), && \text{by Axiom 12.} \\
&= \left[(ab)^{-1}a\right]\left[a^{-1}(bb^{-1})\right], && \text{by Axiom 12.} \\
&= \left[(ab)^{-1}a\right]\left[(a^{-1}b)b^{-1}\right], && \text{by Axiom 12} \\
&= \left[(ab)^{-1}a\right]\left[(ba^{-1})b^{-1}\right], && \text{by Axiom 11} \\
&= \left[(ab)^{-1}a\right]\left[b(a^{-1}b^{-1})\right], && \text{by Axiom 12} \\
&= \left(\left[(ab)^{-1}a\right]b\right)\left(a^{-1}b^{-1}\right), && \text{by Axiom 12} \\
&= \left[(ab)^{-1}(ab)\right](a^{-1}b^{-1}), && \text{by Axiom 12} \\
&= 1(a^{-1}b^{-1}), && \text{by Axiom 14} \\
&= a^{-1}b^{-1}, && \text{by Axiom 13.} \clubsuit
\end{aligned}
$$

Theorem 31:

1. There is one and only one zero element, 0, in $\mathbb{R}$.
2. There is one and only one unit element, 1, in $\mathbb{R}$.

Proof:

31.1. That there is at least one zero element, 0, in $\mathbb{R}$ follows from Axiom 8. Hence, all we need to show is that the zero element of $\mathbb{R}$ is unique. We will use the method of **proof by contradiction.**

Assume there are more than one zero elements. This being the case, there is at least another zero element, say z, distinct from 0. Now, let us calculate the sum $0+z$ in two different ways. First of all, we know that 0 is a zero element of $\mathbb{R}$, thus:

$$0+z=z.$$

That is,

$$z=0+z, \qquad \text{by Axiom 2.} \qquad (1)$$

On the other hand, the assumption that z too is a zero element of $\mathbb{R}$ yields:

$$0+z=0. \qquad (2)$$

Therefore, using Axiom 3, we conclude from (1) and (2) that $z=0$, which contradicts the earlier assumption that z and 0 were distinct.

31.2. The proof of this part of Theorem 31 is very much similar to the proof of part 1. Indeed, assume there are more than one unit elements. This being the case, there is at least another unit element, say u, distinct from 1. Now, let us calculate the product $1u$ in two different ways.

First of all, we know that 1 is a unit element of $\mathbb{R}$, thus:

$$1u = u .$$

On the other hand, the assumption that u too is a unit element of $\mathbb{R}$ yields:

$$1u = 1 .$$

Therefore, by using Axioms 2 and 3, we can conclude from the above that $u = 1$, which contradicts our earlier assumption that u and 1 were distinct . ♣

Theorem 32:

1. Every element $a \in \mathbb{R}$ has one and only one opposite [or additive inverse], $-a$, in $\mathbb{R}$.
2. Every **non zero** element $a \in \mathbb{R}$ has one and only one reciprocal [or multiplicative inverse], a^{-1} , in $\mathbb{R}$.

Proof:

32.1. That every real number a has at least one opposite follows from Axiom 9. Therefore, all we need to show is that given any real number, its opposite is unique. We will use the method of **proof by contradiction.**

So let a be any given real number. Suppose it has at least two distinct opposites, say a' and a'' . We will calculate the sum $[a' + a] + a''$ in two different ways:

On the one hand, we have:

$$[a' + a] + a'' = 0 + a'' , \qquad \text{by Axiom 9}$$

$$= a'' , \qquad \text{by Axiom 8}$$

that is,

$$[a' + a] + a'' = a''. \tag{3}$$

On the other hand

$$\begin{aligned} [a' + a] + a'' &= a' + [a + a''] && \text{by Axiom 7} \\ &= a' + 0 && \text{by Axiom 9} \\ &= a', && \text{by Axiom 8} \end{aligned}$$

which gives

$$[a' + a] + a'' = a'. \tag{4}$$

Therefore, using (3) and (4) together with Axioms 2 and 3, we get $a' = a''$, which contradicts our earlier assumption that a' and a'' were distinct elements.

32.2. The proof of part 2 of this theorem is similar to that of part 1. All one needs to do to prove it is to replace in the above, on the one hand, addition " + " by multiplication " • ", and on the other hand, 0 by 1. So, we urge the reader to supply the details of the proof to this part. ♣

To close this section we present one more theorem, the so-called **Zero Factor Theorem**. This theorem is often invoked in solving quadratic equations.

Theorem 33 [The Zero Factor Theorem]: let a and b be real numbers.

$$ab = 0 \quad \Rightarrow \quad a = 0 \ \textbf{or} \ b = 0.$$

Proof by exhaustion of all cases: There are two cases: $a = 0$ or $a \neq 0$. Hence, we shall use the method of proof by exhaustion.

Case 1. $a = 0$. In this case there is nothing more to prove; since the conjunction $a = 0$ or $b = 0$ is then automatically true; [see Problem 12 of Chapter 2.] This is an example of **the method of trivial proofs**.

Case 2. $a \neq 0$. Then by Axiom 14, a possesses a reciprocal $a^{-1} \in \mathbb{R}$. Thus, by multiplying both sides of the equation $ab = 0$, we have:

$$\begin{aligned} ab = 0 &\Rightarrow a^{-1}(ab) = a^{-1}0, && \text{by Axiom 4} \\ &\Rightarrow a^{-1}(ab) = 0, && \text{by Theorem 22} \\ &\Rightarrow (a^{-1}a)b = 0, && \text{by Axiom 12} \\ &\Rightarrow 1b = 0, && \text{by Axiom 14} \\ &\Rightarrow b = 0, && \text{by Axiom 13} \\ &\Rightarrow a = 0 \text{ or } b = 0. && \text{by Problem 12 of Chapter 2.} \end{aligned}$$

Hence, in either case, we arrive at $a = 0$ or $b = 0$, which completes the proof. ♣

Example 34. Solve the quadratic equation: $x^2 - x - 6 = 0$.

Solution

We know that

$$x^2 - x - 6 = (x - 3)(x + 2).$$

Therefore, by Axiom 4,

$$x^2 - x - 6 = 0 \quad \Leftrightarrow \quad (x - 3)(x + 2) = 0$$

But this too is equivalent to

$$x - 3 = 0 \textbf{ or } x - 2 = 0, \qquad \text{by Theorem 33}$$

which in turn is equivalent to

$$x = 3 \quad \textbf{or} \quad x = 2 \,. \heartsuit$$

9.3. SOME CONSEQUENCES OF THE ORDER AXIOMS

In the last section we discussed some of the properties of real numbers involving just the addition and multiplication operations. In the present section, we propose to go further; we will examine a few properties involving not just these two operations but also the order relation " $<$."

But before doing that, we will need the following

Definition 35:

1. A real number x is said to be **negative**, if and only if $x < 0$.
2. A real number x is said to be **positive**, if and only if $0 < x$.
3. A real number x is said to be **greater than** another real number y, in which case we write $x > y$, if and only if $y < x$.

We are now in a position to examine some of the consequences of the order axioms on real numbers.

Theorem 36: Let $a \in \mathbb{R}$.

1. If $a < 0$, then $-a > 0$.
2. If $a > 0$, then $-a < 0$.

Proof:

36.1. Suppose $a < 0$.

$a < 0$	$\Rightarrow$	$a + (-a) < 0 + (-a)$	by Axiom 18
	$\Rightarrow$	$a + (-a) < -a$	by Axiom 8
	$\Rightarrow$	$0 < -a$	by Axiom 9
	$\Rightarrow$	$-a > 0$	by Definition 35.3

36.2. The proof of part 2 of this theorem is almost identical to that of part 1. So, we urge the reader to supply the details of the proof for this part. ♣

Definition 37:

1. A real number x is said to be **less than, or equal to** a real number y, in which case we write $x \leq y$, if and only if $x < y$ or $x = y$.

2. A real number x is said to be **greater than, or equal to** a real number y, in which case we write $x \geq y$, if and only if $x > y$ or $x = y$.

Exercise 38: Let x and y be real numbers. Then show that

1. $x \leq y \iff y \geq x$
2. $(x \leq y \text{ and } y \leq x) \iff x = y.$
3. $(x \geq y \text{ and } y \geq x) \iff x = y.$

Solution:

38.1. $x \leq y \iff x < y$ or $x = y$, by Definition 37.1

$\iff y > x$ or $x = y$, by Definition 35.3

$\iff y \geq x$, by Definition 37.2

38.2. We know that

$$(x \leq y \text{ and } y \leq x)$$

$$\Updownarrow$$

$$\Big[(x < y) \vee (x = y)\Big] \wedge \Big[(y < x) \vee (y = x)\Big],$$

which, by the results of Problem 16 of Chapter 2,

$$\Leftrightarrow \begin{cases} x < y, \ \ and \ \ y < x, \\ or \\ x < y, \ \ and \ \ y = x, \\ or \\ x = y, \ \ and \ \ y < x, \\ or \\ x = y, \ \ and \ \ y = x \end{cases}$$

which by Axiom 16, gives

$$\Leftrightarrow \quad CONT \vee CONT \vee CONT \vee (x = y)$$

which, by the result of Problem 16 of Chapter 2,

$$\Leftrightarrow \quad x = y.$$

38.3. Clearly, using Parts 1 and 2 of Exercise 38, we have:

$$\begin{aligned} (\ x \geq y \ \text{ and } \ y \geq x\ ) &\Leftrightarrow (\ y \leq x \ \text{ and } \ x \leq y\ ) \\ &\Leftrightarrow (\ x \leq y \ \text{ and } \ y \leq x\ ) \\ &\Leftrightarrow x = y. \ \heartsuit \end{aligned}$$

Theorem 39: Let a, x, and y be any three real numbers. Then

1. $x < y$ and $a < 0 \Rightarrow ax > ay$.
2. $a \neq 0 \Rightarrow a^2 > 0$.
3. $a > 0 \Rightarrow a^{-1} > 0$.
4. $a < 0 \Rightarrow a^{-1} < 0$.
5. $0 < x < y \Rightarrow 0 < y^{-1} < x^{-1}$.

Proof:

39.1. We know, by Theorem 36.1, that

$$(x < y \text{ and } a < 0)$$

$$\Downarrow$$

$$(x < y \text{ and } -a > 0)$$

$$\Rightarrow (-a)x < (-a)y \quad \text{by Axiom 19}$$

$$\Rightarrow -(ax) < -(ay) \quad \text{by Theorem 26}$$

$$\Rightarrow -(ax) + (ay) < -(ay) + (ay) \quad \text{by Axiom 18}$$

$$\Rightarrow -(ax) + (ay) < 0 \quad \text{by Axiom 9}$$

$$\Rightarrow -\left[-(ax) + (ay)\right] > 0 \quad \text{by Theorem 36.1}$$

$$\Rightarrow (-1)\left[-(ax) + (ay)\right] > 0 \quad \text{by Theorem 24}$$

$$\Rightarrow (-1)\left[-(ax)\right] + (-1)(ay) > 0 \quad \text{by Axiom15}$$

$$\Rightarrow ax + [-(ay)] > 0 \quad \text{by Theorems 27 and 24}$$

$$\Rightarrow ax + [-(ay)] + ay > ay \quad \text{by Axiom18}$$

$$\Rightarrow ax > ay \quad \text{by Axioms 7, 9 and 8}$$

39.2. Since $a \neq 0$, by Axiom16, we know that there remain all in all two cases here: $0 < a$ and $a < 0$. Hence, we shall use **the method of proof by exhaustion of all cases.**

Case 1:

$$0 < a \Rightarrow 0a < aa \quad \text{by Axiom 19}$$

$$\Rightarrow 0 < aa \quad \text{by Theorem 22}$$

$$\Rightarrow \quad 0 < a^2 .$$

Case 2:

$$a < 0 \Rightarrow 0 < -a \quad \text{by Theorem 36.1}$$
$$\Rightarrow 0(-a) < (-a)(-a) \quad \text{by Axiom 19}$$
$$\Rightarrow 0 < (-a)(-a) \quad \text{by Theorem 22}$$
$$\Rightarrow 0 < a^2 . \quad \text{by Theorem 27}$$

Hence, in either case, we do arrive at $0 < a^2$, which is the same thing as $a^2 > 0$. Thus the proof of Part 2 is complete.

39.3. This part of the theorem may be proved easily by **contradiction**. But first, let us observe that $1 > 0$. Indeed, using part 2 of this theorem, together with Axiom13, we immediately discover that $1 = 1^2 > 0$. Now, let us assume that the implication $a > 0 \Rightarrow a^{-1} > 0$ is false. This would mean that there is at least one $a > 0$ whose reciprocal $a^{-1} \leq 0$. Hence, applying Axiom19, we have $a^{-1}a \leq 0a$. That is $1 \leq 0$, which contradicts the fact that $1 > 0$.

39.4. The proof of this part of the theorem is almost identical to that of part 3. So we leave it to the reader to supply all the details.

39.5. Let us assume that $0 < x < y$. Then by part 3 of this theorem, both $0 < x^{-1}$ and $0 < y^{-1}$ are true. Therefore, using Axiom19, we have:

$$x < y \Rightarrow y^{-1}x < y^{-1}y$$
$$\Rightarrow y^{-1}x < 1 \quad \text{by Axiom 14}$$
$$\Rightarrow (y^{-1}x)x^{-1} < x^{-1} \quad \text{by Axioms 19 and 13}$$

$$\Rightarrow \quad y^{-1}(xx^{-1}) < x^{-1} \quad \text{by Axiom 12}$$
$$\Rightarrow \quad y^{-1} \cdot 1 < x^{-1} \quad \text{by Axiom 14}$$
$$\Rightarrow \quad y^{-1} < x^{-1} \quad \text{by Axioms 13.} \;\clubsuit$$

PROBLEMS

1. Let x, y and z be any three real numbers. Prove that

$$\text{if } xz = yz \text{ and } z \neq 0, \text{ then } x = y.$$

2. Let x, y and z be real any three numbers. Prove that

$$\text{if } x + z = y + z, \text{ then } x = y.$$

3. Let G be a nonempty set. Furthermore, assume a binary operation $\otimes$, such as addition or multiplication is defined on pairs of elements of G. The system $(G, \otimes)$, made up of the set G together with the operation $\otimes$, is called a **group** provided that the following four conditions hold true:

C.1. If $a \in G$ and $b \in G$, then $a \otimes b$ is a unique element of G.
C.2. If $a \in G$, $b \in G$, and $c \in G$, then $(a \otimes b) \otimes c = a \otimes (b \otimes c)$.
C.3. There is a fixed element of G, say e and called the *identity* element in G, such that

$$a \otimes e = e \otimes a = a, \quad \text{for every } a \in G.$$

C.4. For each $a \in G$, there is an element of G called the inverse of a and denoted by a^{-1}, such that

$$a \otimes a^{-1} = a^{-1} \otimes a = e.$$

3.1. Verify that if $G = \{1, -1\}$ and $\otimes$ is the ordinary multiplication, then $(G, \otimes)$ is a group.

3.2. Verify that if $G = \{a + b\sqrt{2} : a \in \mathbb{Q} \text{ and } b \in \mathbb{Q}\}$ and $\otimes$ is the ordinary addition, then $(G, \otimes)$ is a group.

3.3. Verify that if $G = \{a + b\sqrt{2} \neq 0 : \ a \in \mathbb{Q} \ and \ b \in \mathbb{Q}\}$ and $\otimes$ is the ordinary multiplication, then $(G, \otimes)$ is a group.

4. Prove that every group has one and only one identity element.

5. Let x, y and z be elements of a group. Prove that

$$\text{if } x \otimes z = y \otimes z, \text{ then } x = y.$$

6. Prove that each element of a group has a unique inverse.

7. Let a and b be elements of a group $(G, \otimes)$. Prove that the inverse of $a \otimes b$ is $b^{-1} \otimes a^{-1}$.

8. Let $(G, \otimes)$ be a group and let a and b be any two elements of G. Then prove:

$$(a \otimes b)^2 = a^2 \otimes b^2 \Leftrightarrow a \otimes b = b \otimes a.$$

9. * Prove that the multiplicative inverse of the real number 1 is 1 itself.

10. Let x, y and z be any three real numbers. Show that

$$\text{if } x = y, \text{ then } xz = yz.$$

11. From the definition of a group, Condition $C1$, can you easily see why no group can be empty?

"... algebra is the intellectual instrument which has been created for rendering clear the quantitative aspects of the world."
Alfred North Whitehead (1861-1947)

CHAPTER 10

REAL NUMBERS
SUBTRACTION AND DIVISION

In this chapter we will define the operations of subtraction and division on real numbers. We will also study some of the properties of these two operations.

10.1. SUBTRACTION

Definition 1 [Subtraction]: If a and b are in $\mathbb{R}$, then

$$a-b=a+(-b).$$

In other words, to subtract b from a is the same thing as to add the opposite of b to a.

Example 2: Perform the following subtractions

1. $6-9$, 2. $(-5)-9$, 3. $(-5)-(-3)$

Solution

2.1. $6-9=6+(-9)=-3$.

2.2. $(-5)-9=(-5)+(-9)=-14$.

3.3. $(-5)-(-3)=(-5)+\left[-(-3)\right]=(-5)+3=-2$

. ♥

Remark 3: As you might have noticed, we are using the sign "$-$" here in two different ways. In some cases, it indicates the operation of subtraction, as in $6-9$; in other cases, it indicates the opposite of a real number, as in -5, or $-(-3)$. Thus, the actual meaning is generally clear from the context. ■

From the foregoing definition, we immediately obtain the following

Theorem 4: If a is in $\mathbb{R}$, then $0-a=-a$.

Proof:

$$0-a = 0+(-a) \quad \text{by Definition 1}$$

$$= -a \quad \text{by Axiom 8 of Chapter 9.} ♣$$

Example 5: Simplify the following expressions

1. $0-3$
2. $0-3x^2+5$

3. $0 - 3(x^2 + 3x)$

Solution:

5.1. $0 - 3 = -3$

5.2. $0 - 3x^2 + 5 = [0 - 3x^2] + 5 = -3x^2 + 5$

5.3. $0 - 3(x^2 + 3x) = 0 - [3(x^2 + 3x)] = -3(x^2 + 3x)$

. ♥

The next theorem is an extension of the concept of **distributivity** to subtraction. We may therefore call it the **distributive property of multiplication over subtraction**.

Theorem 6: If a, b, and c are in $\mathbb{R}$, then $a(b-c) = ab - ac$, and $(b-c)a = ba - ca$,

Proof:

$$a(b-c) = a[b + (-c)] \quad \text{by Definition 1}$$
$$= ab + a(-c) \quad \text{by Axiom 15 of Chapter 9}$$
$$= ab + [-ac] \quad \text{by Theorem 26 of Chapter 9}$$
$$= ab - ac \quad \text{by Definition 1.}$$ ♣

Example 7: Expand and simplify these expressions

1. $2(5x - 4y)$
2. $-3[5x - 4y]$
3. $10 - 3[5x - 4y]$

Solution:

7.1. $2(5x - 4y) = 2(5x) - 2(4y)$ by Theorem 6

$= (2 \cdot 5)x - (2 \cdot 4)y$ by Axiom 12 of Chapter 9

$= 10x - 8y$

7.2. $-3[5-4y] = (-3)[5x-4y]$ by Theorem 26 of Chapter 9

$= (-3)(5x) - (-3)(4y)$ by Theorem 6

$= (-15x) - (-12y)$ by Theorem26 of Chapter 9

$= -15x + 12y$ by Definition 1

7.3. $10 - 3[5-4y] = 10 + (-3[5-4y])$ by Definition 1

$= 10 + ([-3][5-4y])$ by Theorem 26 of Chapter 9

$= 10 + ([-3][5] - [-3][4y])$ by Theorem 6

$= 10 + ([-15] - [-12y])$ by Theorem 26 of Chapter 9

$= 10 + (-15 + 12y)$ by Definition 1

$= -5 + 12y$. ♥

Exercise 8: show that the following equalities hold for all a and b in $\mathbb{R}$.

1. $-(a+b) = -a - b$
2. $-(a-b) = -a + b$

Solution:

8.1.

$$
\begin{aligned}
-[a+b] &= (-1)[a+b] && \text{by Theorem 24 of Chapter 9} \\
&= (-1)a + (-1)b && \text{by Axiom 15 of Chapter 9} \\
&= (-a) + (-b) && \text{by Theorem 24 of Chapter 9} \\
&= -a - b && \text{by Definition 1}
\end{aligned}
$$

8.2.

$$
\begin{aligned}
-[a-b] &= (-1)[a-b] && \text{by Theorem 24 of Chapter 9} \\
&= (-1)a - (-1)b && \text{by Theorem 6} \\
&= (-a) - (-b) && \text{by Theorem 24 of Chapter 9} \\
&= -a + b && \text{by Definition 1.}
\end{aligned}
$$

♥

10.2. DIVISION

We now introduce the concept of division of a real number by a real number.

Definition 9 [Division]: If a and b are in $\mathbb{R}$, then

$$\frac{a}{b} = a\,b^{-1}, \quad \text{provided} \quad b \neq 0.$$

In other words, to divide a by b is to multiply a by b^{-1}, provided that $b \neq 0$.

With this definition, we easily prove the following

Theorem 10 If a and b are in $\mathbb{R}$,

1. $\frac{a}{1} = a$

2. $\frac{1}{b} = b^{-1}$ provided $b \neq 0$.

3. $\frac{a}{1} \cdot \frac{1}{b} = \frac{a}{b}$ provided $b \neq 0$.

4. $\frac{a}{a} = 1$ provided $a \neq 0$.

5. $\frac{a}{b} = a \cdot \frac{1}{b}$ provided $b \neq 0$.

Proof:

10.1. $\frac{a}{1} = a \cdot 1^{-1}$ by Definition 9

$= a \cdot 1$ since $1^{-1} = 1$

$= a$ by Axiom 13 of Chapter 9 .

10.2. $\frac{1}{b} = 1 \cdot b^{-1}$ by Definition 9

$= b^{-1}$ by Axiom13 of Chapter9

10.3. $\frac{a}{1} \cdot \frac{1}{b} = a \cdot b^{-1}$ by using parts 1 and 2 of the present theorem.

$= \frac{a}{b}$ by Definition 9.

10.4. $\frac{a}{a} = a \cdot a^{-1}$ by Definition 9

$= 1$ by Axiom14 of Chapter 9.

10.5. This part of the theorem is a immediate consequence of Part 2. Indeed,

$\frac{a}{b} = a \cdot b^{-1}$ by Definition 9

$= a \cdot \frac{1}{b}$ by using part 2 of this theorem. ♣

Example 11: Solve the algebraic equation

$$2(x-3) - \frac{x+3}{5} = -7 \qquad (1)$$

Solution:

First, we multiply both sides of the original equation by 5 so as to transform it into another equation that is equivalent, but involves no fractions:

$(1) \Leftrightarrow \quad 5\left[2(x-3) - \frac{x+3}{5}\right] = 5[-7]$ by Axiom 4 of Chapter 9

$\Leftrightarrow \quad 5[2(x-3)] - 5\left[\frac{x+3}{5}\right] = 5[-7]$ by Theorem 6

$\Leftrightarrow \quad 5[2(x-3)] - 5\left[(x+3)\frac{1}{5}\right] = 5[-7]$ by Theorem 10.5

$\Leftrightarrow \quad 10(x-3) - 5\left[(x+3)\frac{1}{5}\right] = 5[-7]$ by Axiom 12 of Chapter 9

$$\Leftrightarrow \quad 10(x-3)-5\left[(x+3)\frac{1}{5}\right]=-35 \qquad \text{by Theorem 26 of Chapter 9}$$

$$\Leftrightarrow \quad 10(x-3)-5\left[\frac{1}{5}(x+3)\right]=-35 \qquad \text{by Axiom 11 of Chapter 9}$$

$$\Leftrightarrow \quad 10(x-3)-\left[5\frac{1}{5}\right](x+3)=-35 \qquad \text{by Axiom 12 of Chapter 9}$$

$$\Leftrightarrow \quad 10(x-3)-[1](x+3)=-35 \qquad \text{by Theorem 10.4}$$

$$\Leftrightarrow \quad 10(x-3)-(x+3)=-35 \qquad \text{by Axiom 13 of Chapter 9}$$

$$\Leftrightarrow \quad 10x-30-x-3=-35 \qquad \text{by Theorem 6 and Exercise 8}$$

$$\Leftrightarrow \quad 9x-33=-35$$

$$\Leftrightarrow \quad x=-\frac{2}{9}. \ \heartsuit$$

The next theorem teaches us how to multiply two quotients of real numbers.

Theorem 12: If a, b, c and d are in $\mathbb{R}$, then

$$\frac{a}{b}\cdot\frac{c}{d}=\frac{a\cdot c}{b\cdot d}, \quad \text{provided} \quad b\neq 0, \text{ and } \quad d\neq 0.$$

Proof:

$$\frac{a}{b}\cdot\frac{c}{d} = (ab^{-1})\cdot(cd^{-1}) \qquad \text{by Definition 9}$$

$= a\left[b^{-1}(cd^{-1})\right]$ by Axiom 12 of Chapter 9

$= a\left[(b^{-1}c)d^{-1}\right]$ by Axiom 12 of Chapter 9

$= a\left[(cb^{-1})d^{-1}\right]$ by Axiom 11 of Chapter 9

$= a\left[c\left(b^{-1}d^{-1}\right)\right]$ by Axiom 12 of Chapter 9

$= (ac)\left(b^{-1}d^{-1}\right)$ by Axiom 12 of Chapter 9

$= (ac)\cdot(bd)^{-1}$ by Theorem 30 of Chapter 9

$= \dfrac{ac}{bd}$ by Definition 9. ♣

Exercise 13: show that if a and b are both non zero real numbers, then

1. $\dfrac{1}{\frac{a}{b}} = \dfrac{b}{a}$

2. $\dfrac{\frac{1}{a}}{b} = \dfrac{1}{ab}$

Solution:

1. From Theorem 12, it is easy to see that $\frac{a}{b}\cdot\frac{b}{a} = \frac{ab}{ba} = 1$. That is to say, $\frac{a}{b}\cdot\frac{b}{a} = 1$. Therefore, $\frac{a}{b}$ and $\frac{b}{a}$ are reciprocals of one another. Thus,

$$\frac{1}{\frac{a}{b}} = \frac{b}{a}.$$

2. We know from Theorem 10.2 that $\frac{1}{a} = a^{-1}$, thus

$$\frac{\frac{1}{a}}{b} = \frac{a^{-1}}{b}$$

$$= a^{-1}b^{-1} \quad \text{by Definition 9}$$

$$= (ab)^{-1} \quad \text{by Theorem 30 of Chapter 9}$$

$$= \frac{1}{ab} \quad \text{by Theorem 10.2.} \ \heartsuit$$

We now present a theorem that shows us how to divide a quotient of real numbers by another quotient of real numbers.

Theorem 14: If a, b, c and d are in $\mathbb{R}$, then

$$\frac{\frac{a}{b}}{\frac{c}{d}} = \frac{a}{b} \cdot \frac{d}{c}, \quad \text{provided} \quad b \neq 0, \quad c \neq 0, \text{ and } \quad d \neq 0.$$

Proof:

$$\frac{\frac{a}{b}}{\frac{c}{d}} = \frac{\left(\frac{a}{b}\right)}{\left(\frac{c}{d}\right)}$$

$$= \left(\frac{a}{b}\right) \cdot \frac{1}{\left(\frac{c}{d}\right)} \quad \text{by Theorem 10.5}$$

$$= \frac{a}{b} \cdot \frac{d}{c} \quad \text{by the result of Exercise 13.1.} \ \clubsuit$$

Example 15: Calculate the expression $\dfrac{\frac{2}{3}}{\frac{4}{7}}$

Solution

$$\begin{aligned}
\frac{\frac{2}{3}}{\frac{4}{7}} &= \frac{2}{3}\cdot\frac{7}{4} && \text{by Theorem 14} \\
&= \frac{14}{12} && \text{by Theorem 12} \\
&= \frac{7}{6}. \heartsuit
\end{aligned}$$

Our next theorem has such a great importance in algebra that it is customarily referred to in some textbooks as the **Golden Rules of Fractions**.

Theorem 16 [Golden Rules of Fractions]: If a, b, and k are in $\mathbb{R}$, then

1. $\dfrac{ak}{bk} = \dfrac{a}{b}$, provided $b \neq 0$, and $k \neq 0$,
2. $-\dfrac{a}{b} = \dfrac{-a}{b} = \dfrac{a}{-b}$, provided $b \neq 0$.

Proof:

1.
$$\begin{aligned}
\frac{ak}{bk} &= (ak)\cdot(bk)^{-1} && \text{by Definition 9} \\
&= (ak)\cdot(b^{-1}k^{-1}) && \text{by Theorem 30 of Chapter 9} \\
&= (ak)\cdot(k^{-1}b^{-1}) && \text{by Axiom 11 of Chapter 9}
\end{aligned}$$

$= a\left[k(k^{-1}b^{-1})\right]$ by Axiom 12 of Chapter 9

$= a\left[(kk^{-1})b^{-1}\right]$ by Axiom 12 of Chapter 9

$= a\left[(1)b^{-1}\right]$ by Axiom 14 of Chapter 9

$= a\left[b^{-1}\right]$ by Axiom 13 of Chapter 9

$= ab^{-1}$

$= \frac{a}{b}$ by Definition 9

2. $* -\frac{a}{b} = -\left[ab^{-1}\right]$ by Definition 9

$= (-a)b^{-1}$ by Theorem 26 of Chapter 9

$= \frac{-a}{b} *$ by Definition 9

$= (-a)b^{-1}$ by Definition 9

$= a(-b^{-1})$ by Theorem 26 of Chapter 9

$= a\left[(-1)b^{-1}\right]$ by Theorem 24 of Chapter 9

$= a\left[(-1)^{-1}b^{-1}\right]$ since $-1 = (-1)^{-1}$

$= a\left[(-1)b\right]^{-1}$ by Theorem 30 of Chapter 9

$= a\left[-b\right]^{-1}$ by Theorem 24 of chapter 9

$= \frac{a}{-b} *$ by Definition 9. ♣

Example17: Reduce the following fraction to its lowest terms $\frac{18}{24}$

Solution:
First notice that

$$\frac{18}{24} = \frac{3 \cdot 6}{4 \cdot 6}$$

Then using Theorem 16.1, we immediately have

$$= \frac{3}{4} \quad \text{by cancelling out the common factor } 6 \text{ .} ♥$$

Our final theorem below tells us how to add two quotients. The first equality gives us the sum of two quotients that share the same denominator; whereas the second equality is about the sum of two quotients with distinct denominators.

Theorem 18: If a , b and k are in $\mathbb{R}$, then

1. $\frac{a}{k} + \frac{b}{k} = \frac{a+b}{k}$, provided $k \neq 0$,

2. $\frac{a}{b} + \frac{c}{d} = \frac{ad+bc}{bd}$, provided $b \neq 0$, and $d \neq 0$.

Proof:

1. $\frac{a}{k} + \frac{b}{k} = ak^{-1} + bk^{-1}$ by Definition 9

$= (a+b)k^{-1}$ by Axiom15 of Chapter 9

$= \frac{a+b}{k}$ by Definition 9

2. $\frac{a}{b} + \frac{c}{d} = ab^{-1} + cd^{-1}$ by Definition 9

$= (ab^{-1})[1] + (cd^{-1})[1]$ by Axiom 13 of Chapter 9

$= (ab^{-1})[(bd)(bd)^{-1}] + (cd^{-1})[(bd)(bd)^{-1}]$ by Axiom 14 of Chapter 9

$= [(ab^{-1})(bd)](bd)^{-1} + [(cd^{-1})(bd)](bd)^{-1}$ by Axiom 12 of Chapter 9

$= [\,(ab^{-1})(bd) + (cd^{-1})(bd)\,](bd)^{-1}$ by Axiom 15 of Chapter 9

$= [\,a[b^{-1}(bd)] + c[d^{-1}(bd)]\,](bd)^{-1}$ by Axiom 12 of Chapter 9

$= [\,a[b^{-1}(bd)] + c[d^{-1}(db)]\,](bd)^{-1}$ by Axiom 11 of Chapter 9

$= [\,a[(b^{-1}b)d] + c[(d^{-1}d)b]\,](bd)^{-1}$ by Axiom 12 of Chapter 9

$= [\,a[(1)d] + c[(1)b]\,](bd)^{-1}$ by Axiom 14 of Chapter 9

$= [\,a[d] + c[b]\,](bd)^{-1}$ by Axiom 13 of Chapter 9

$= [\,ad + cb\,](bd)^{-1}$

$= \frac{ad + bc}{bd}$ by Definition 9. ♣

Example 19: Calculate the following expressions

1. $\dfrac{3x-1}{2x}+\dfrac{2-7x}{2x}$, where $x \neq 0$.

2. $\dfrac{3x-1}{2x-1}+\dfrac{2-7x}{5-2x}$, where $x \neq \dfrac{1}{2}$, and $x \neq \dfrac{5}{2}$.

Solution:

1. Since both fractions share the same denominator, we use Part 1 of Theorem 18:

$$\frac{3x-1}{2x}+\frac{2-7x}{2x} = \frac{(3x-1)+(2-7x)}{2x} = \frac{-4x+1}{2x}, \quad x \neq 0.$$

2. Since the two fractions do not share the same denominator, we use Part 2 of Theorem 18:

$$\frac{3x-1}{2x-1}+\frac{2-7x}{5-2x} = \frac{(3x-1)(5-2x)+(2-7x)(2x-1)}{(2x-1)(5-2x)}$$

$$= \frac{-20x^2+28x-7}{(2x-1)(5-2x)}, \quad \text{where } x \neq \frac{1}{2}, \text{ and } x \neq \frac{5}{2}. \ ♥$$

PROBLEMS

1. Let a, b, $c \neq 0$ and $d \neq 0$ be real numbers. Prove that

if $\dfrac{a}{b}=\dfrac{c}{d}$ and $b \neq -d$, then $\dfrac{a}{b}=\dfrac{a+c}{b+d}=\dfrac{c}{d}$.

2. Let a, b, $c \neq 0$ and $d \neq 0$ be real numbers. Prove that

if $\frac{a}{b} = \frac{c}{d}$ and $b \neq d$, then $\frac{a}{b} = \frac{a-c}{b-d} = \frac{c}{d}$.

3. Solve the following system of equations using the result of Problem 1:

$$\begin{cases} \frac{x}{2} = \frac{2y}{5} \\ 2x + 4y = 10 \end{cases}$$

4. Solve the following system of equations using the result of Problem 2:

$$\begin{cases} \frac{x}{2} = \frac{2y}{5} \\ 2x - 4y = 10 \end{cases}$$

5. Assume $a, b, x,$ and y are real numbers. Then show that:

$$a = b \text{ and } x = y \Rightarrow a + x = b + y.$$

6. * Assume $a, b, x,$ and y are real numbers. Then show that:

$$a \leq b \text{ and } x \leq y \Rightarrow a + x \leq b + y.$$

"Pure mathematics is, in its way, the poetry of logical ideas."
Albert Einstein.

CHAPTER 11

REAL NUMBERS
THE AXIOM OF CONTINUITY

In this chapter we will discuss the **axiom of continuity** mentioned in Chapter 9. As will be seen, this is a fundamental axiom, in that many properties of the real numbers are based on it. Thus, a sound understanding of this axiom is important for anyone contemplating a rigorous study of the calculus or other advanced topics in mathematics.

However, before presenting the axiom of continuity and exploring its consequences, we will need to first introduce a few elementary notions. This is done in the next two sections.

11.1. SOME SUBSETS OF THE SET OF REAL NUMBERS

We are all familiar with the counting numbers. A counting number is also called a **natural number**. The set of natural numbers, often denoted by $\mathbb{N}$, can be described as

$$\mathbb{N} = \{ 1, 2, 3, \ldots \}.$$

Having introduced the natural numbers, we also get, $\mathbb{Z}$, the set of integers which is simply the collection of all natural numbers together with zero and all the opposites of the natural numbers:

$$\mathbb{Z} = \{\ldots, -3, -2, -1, 0, 1, 2, 3, \ldots\}.$$

Another set often encountered in mathematics is the set of **rational numbers** defined by:

$$\mathbb{Q} = \left\{ x \in \mathbb{R} : \quad x = \frac{a}{b}, \quad \text{for some } a \in \mathbb{Z}, \text{ and some } b \in \mathbb{N} \right\}.$$

In other words, a rational number is any real number that can be expressed as the quotient of an integer by a natural number. Hence, as seen in Example 5 of Chapter 3, the number $\sqrt{2}$ is not a rational number. We call any real number, such as $\sqrt{2}$, which is not a rational number an **irrational number**. Therefore, the set of all irrational numbers, denoted by $\mathbb{I}$, is simply the complement of set $\mathbb{Q}$ with respect to $\mathbb{R}$:

$$\mathbb{I} = \mathbb{Q}'.$$

11.2. INTERVALS OF REAL NUMBERS

In addition to the above sets, there are other subsets of $\mathbb{R}$ that are encountered so often in mathematics that it is convenient to give them not only special names but also special notations. One such type of sets are the **intervals**.

Intuitively, an interval of real numbers is any **continuous portion** of the number line. In other words, an interval may be thought of

1. as the entire segment between any two fixed points of the number line
2. or as the entire portion of the number line lying either to the right or to the left of a fixed point .
3. or as the entire number line itself.

In the table below are gathered those types of intervals that are most often met in algebra. Assuming a and b to be fixed real numbers such that $a < b$, in the middle column of the table we give the intervals, and in the right column we give the corresponding descriptions of these intervals in terms of the set-builder notation. And in the left column appear the corresponding names of these intervals.

Name	Interval notation	Set-builder notation
Open interval	$]a, b[$	$\{x \in \mathbb{R}: \ a < x < b\}$
Closed interval	$[a, b]$	$\{x \in \mathbb{R}: \ a \le x \le b\}$
Left-closed, right-open interval	$[a, b[$	$\{x \in \mathbb{R}: \ a \le x < b\}$
Left-open, right-closed interval	$]a, b]$	$\{x \in \mathbb{R}: \ a < x \le b\}$
Left-open, plus-infinity interval	$]a, \infty[$	$\{x \in \mathbb{R}: \ x > a\}$
Left-closed, plus-infinity interval	$[a, \infty[$	$\{x \in \mathbb{R}: \ x \ge a\}$
Minus-infinite, right-open interval	$]-\infty, a[$	$\{x \in \mathbb{R}: \ x < a\}$
Minus-infinite, right-closed interval	$]-\infty, a]$	$\{x \in \mathbb{R}: \ x \le a\}$
$\mathbb{R}$	$]-\infty, \infty[$	$\{x \in \mathbb{R}: \ x \in \mathbb{R}\}$

Remark 1: Note that none of the above intervals has been closed at $-\infty$ or ∞. Indeed, since an interval is by definition a subset of $\mathbb{R}$, closing it at either $-\infty$ or ∞ would suggest that $-\infty$ or ∞ is a real number, which is not the case. The symbols $-\infty$ and ∞ are simply convenient notations for minus and plus infinities; they are not real numbers. ■

11.3. THE AXIOM OF CONTINUITY

STATEMENT OF THE AXIOM OF CONTINUITY

You are already familiar with the depiction of the set $\mathbb{R}$, of real numbers, as a continuous line where every point is the graph of a unique real number, and every real number the coordinate of a unique point on the line. As we shall find out, this picture of the real numbers as a "continuous" line, with no "gaps" in it, is a powerful concept from which many important consequences will be deduced.

Intuitively, it is this view, that the set of real numbers can be thought of as a continuous line, that mathematicians refer to as the **axiom of continuity**.

But if we want this intuitive notion of continuity to be of any use, we must re-formulate it more precisely. We do this as follows

> **Axiom 1 [of continuity]:**
>
> For any two subsets Γ and Σ of the set $\mathbb{R}$, if every element of Γ is less than or equal to any element of Σ, then there exists a fixed real number, say c, such that every element of Γ is less than or equal to c, and c is less than or equal to every element of Σ.

Notice that by using logical symbolism we may re-phrase the above axiom even more precisely as follows

> **Axiom 2 [of continuity]:** $\forall \Gamma \subset \mathbb{R}, \quad \forall \Sigma \subset \mathbb{R},$
>
> if
>
> $$\forall x \in \Gamma, \ \forall y \in \Sigma, \quad x \le y$$
>
> then
>
> $$\exists c \in \mathbb{R}, \ \forall x \in \Gamma, \ \forall y \in \Sigma, \quad x \le c \le y.$$

Clearly, as axioms, the foregoing statements require no proof. They are simply the last of the twenty properties we have selected [in Chapter 9] to define the real numbers system.

Hence we need not prove the Continuity Axiom for $\mathbb{R}$. What we will be doing instead in the remainder of this section is to show via a number of exercises that no such axiom exists for the rational numbers. Consequently, it is impossible to associate with all points of the number-line rational coordinates in a one-to-correspondence. For indeed, some points or "gaps" will remain unpaired whose coordinates will require irrational numbers.

LACK OF CONTINUITY OF RATIONAL NUMBERS

What we mean by a **lack of continuity** in $\mathbb{Q}$ is the fact that Axiom 2 above no longer holds, if we replace everywhere in it the set $\mathbb{R}$ by the set $\mathbb{Q}$. In other words, the following proposition is false

Proposition 3: $\forall \Gamma \subset \mathbb{Q}, \quad \forall \Sigma \subset \mathbb{Q},$

if

$$\forall x \in \Gamma, \quad \forall y \in \Sigma, \quad x \leq y,$$

then

$$\exists c \in \mathbb{Q}, \quad \forall x \in \Gamma, \quad \forall y \in \Sigma, \quad x \leq c \leq y.$$

But, to prove the falsity of Proposition 3 is the same thing as to prove the truth of its denial given below:

Theorem 4: $\exists \Gamma \subset \mathbb{Q}, \quad \exists \Sigma \subset \mathbb{Q},$

such that

$$\forall x \in \Gamma, \quad \forall y \in \Sigma, \quad x \leq y, \tag{1}$$

and

$$\sim (\exists c \in \mathbb{Q}, \quad \forall x \in \Gamma, \quad \forall y \in \Sigma, \quad x \leq c \leq y). \tag{2}$$

Thus we are now in the position to show that the Axiom of Continuity does not hold for the set of rationals. As we have just argued, this is exactly the same as proving Theorem 4. We provide the proof of this in the next three exercises. We will utilized the **method of proof by construction**: The goal will be to construct two nonempty subsets of $\mathbb{Q}$ that satisfies Theorem 4.

Exercise 5: let $A = \{r \in \mathbb{Q}^+ : r^2 < 2\}$ and $B = \{r \in \mathbb{Q}^+ : r^2 > 2\}$, where by $\mathbb{Q}^+$ we are denoting the set of all positive rational numbers. Prove that:

1. Sets A and B are nonempty.
2. Sets A and B are disjoint.
3. Any positive rational number belongs to either set A or set B, but never to both.
4. Every element of A is smaller than any element of B

Solution

5.1. It is easy to see that $1 \in A$, and $2 \in B$; indeed, $1^2 < 2$, and $2^2 > 2$. Thus both of sets A and B are nonempty.

5.2. We can prove this part **by contradiction**. Indeed, assume sets A and B are not disjoint. Then both sets have in common at least one element, say r. Thus, r must satisfy the conditions

$$r^2 < 2 \quad \text{and} \quad r^2 > 2,$$

which contradicts Axiom 16 of Chapter 9 [Axiom of Trichotomy].

5.3. For every positive rational number r, we know that r^2 is also a real number. Thus, according to the Axiom of Trichotomy, one and only one of these relationships holds:

$$r^2 < 2 \quad \text{or} \quad r^2 = 2, \text{ or} \quad r^2 > 2.$$

But, the fact that r^2 is rational excludes the possibility that $r^2 = 2$. Hence, either $r^2 < 2$ or $r^2 > 2$ but not both, which is the same as saying that r belongs to either set A or set B but not to both.

5.4. Let r be any element of set A, and let s be any element of set B. Then

$$r^2 < 2 < s^2.$$

Therefore,

$$r^2 - s^2 < 0,$$

or

$$[r - s][r + s] < 0.$$

But, since $r + s > 0$, the above implies $r - s < 0$, that is to say $r < s$. ♥

We shall now take advantage of the next exercise to introduce yet another example of the **method of proof by construction**. As we said in Chapter 3, this method is based on the very simple idea that to prove the existence of a mathematical object, it suffices (i) to supply a candidate for that object, and (ii) to check to make sure that our candidate object, as constructed, has the desired properties.

Exercise 6: Prove the following statements.

1. Set A has no biggest element. [In other words, for every element r in A, there exists another element s in A bigger than r.]
2. Set B has no smallest element. [In other words, for every element r in B, there exists another element of B, say s, smaller than r.]

Solution:

6.1. *Step one*. For every r in A, let us construct our candidate object s :

We know that both s and r must be in A, and that s must be greater than r. Therefore, their difference, which we denote by d, must satisfy:

$$0 < s - r = d \in \mathbb{Q}^+$$

Hence,

$$\begin{aligned} 0 < s^2 &= [r + d]^2 \\ &= r^2 + 2rd + d^2 \\ &= r^2 + [2r + d]d \\ &= r^2 + [r + s]d < 2 . \end{aligned} \qquad (3)$$

which in turn requires that d satisfies the inequality

$$0 < d < \frac{2 - r^2}{r + s} . \qquad (4)$$

Now, we know that $0 < s < 2$. [Otherwise, $s^2 \geq 2$, and so s would not be in A.] Thus, by choosing

$$d = \frac{2 - r^2}{r + 2}, \qquad (5)$$

we have also chosen a d that fulfills inequality (4). Therefore, for the choice of s, we try

$$s = r + d = r + \frac{2 - r^2}{r + 2} . \qquad (6)$$

Step two. Now that we have constructed a candidate s, to finish the proof we only need to check that s as constructed in Step one and given by formula (6) is indeed greater than r and is in set A:

On the one hand, since $r \in A$, we have $2 - r^2 > 0$. Therefore, (6) implies

$$s = r + \frac{2 - r^2}{r + 2} > r .$$

That is, s is greater than r.

On the other hand, by squaring s, we obtain

$$s^2 = \left[r + \frac{2 - r^2}{r+2}\right]^2 = \left[\frac{2r+2}{r+2}\right]^2$$

which, by the fact that $r^2 - 2 < 0$, gives

$$s^2 - 2 = \left[\frac{2r+2}{r+2}\right]^2 - 2 = \frac{2(r^2-2)}{(r+2)^2} < 0$$

Thus, $s^2 < 2$, and so s too is in set A.

6.2. As for this part of the exercise, we may still use the same expression for s:

$$s = r - \frac{r^2 - 2}{r+2}$$

But here we must show in *Step one* that s is less than r, and in *Step two* that s too is an element of set B. Since the proof of this question is very similar to that of Question 6.1, we leave it to the reader to supply all the details. ♥

So far, what we have managed to prove via Exercises 5 and 6 is only the first part of Theorem 4, which is that there are two nonempty subsets of $\mathbb{Q}$, namely A and B, such that

$$\forall x \in A, \quad \forall y \in B, \qquad x \leq y .$$

Therefore, to complete the proof of Theorem 4 we need to also prove the denial of:

$$\exists c \in \mathbb{Q}, \quad \forall x \in A, \quad \forall y \in B, \qquad x \leq c \leq y .$$

We present the proof of this last part of Theorem 4 as the solution to the following.

Exercise 7: Show that there is **no fixed** rational number c such that, for all $x \in A$ and for all $y \in B$,

$$x \leq c \leq y .$$

Solution:
We will use a combination of strategies: **Proof by contradiction** and **proof by exhaution**. Suppose the proposition we seek to show is false. Then, this means there is at least one fixed rational number c such that:

$$\forall x \in A, \quad \forall y \in B, \quad x \leq c \leq y. \qquad (7)$$

But then, (7) would implies that c is greater than or equal to every element of A. Hence, $c > 0$. On the other hand, we know from Exercise 5 that every positive rational must belong to either set A or set B, but not both. Hence there are two cases:

Case 1. Assume $c \in A$: Since A has no biggest element, there must be at least one element of A, say x_0, such that $c < x_0$, which contradicts the assumption in (7).

Case 2. Assume $c \in B$: Since B has no smallest element, there must be at least one element of B, say y_0, such that $y_0 < c$, which also contradicts the assumption in (7).

Thus, in either case, we reach a contradiction. Hence, the proof is complete. ♥

11.4. LEAST UPPER BOUND PRINCIPLE

In this section we will study some of the properties of real numbers based on the axiom of continuity. The culmination of this section is a theorem called the least upper bound principle. [It is a very important theorem of advanced mathematics.]

We will need the following definitions in order to state this principle properly.

Definition 8 [Bounded Sets]: A nonempty subset Σ of $\mathbb{R}$ is said to **have an upper bound** or to be **bounded from above**, if

$$\exists M \in \mathbb{R}, \quad \forall x \in \Sigma, \quad x \leq M. \tag{8}$$

Definition 9 [Upper Bounds]: Let M be a fixed real number, and Σ a subset of the set of real numbers. Then M is called an **upper bound** of Σ, if

$$\forall x \in \Sigma, \quad x \leq M. \tag{9}$$

Exercise 10: Translate Definitions 8 and 9 into plain English.

Solution

- A nonempty subset Σ of $\mathbb{R}$ is said to have an upper bound or to be bounded from above, if there is a fixed number such that all elements of Σ are less than or equal to that fixed number.
- A fixed number M is called an upper bound of a subset Σ of the set of real numbers, if all elements of Σ are less than or equal to the fixed number M. ♥

Exercise 11: Write the denial of the foregoing definitions

Solution

- *The denial of Definition 8 is*: A nonempty subset Σ of $\mathbb{R}$ is said to **have no upper bound** or to **be unbounded from above**, if

  $$\sim (\exists M \in \mathbb{R}, \quad \forall x \in \Sigma, \quad x \leq M),$$

  that is if

  $$\forall M \in \mathbb{R}, \quad \exists x \in \Sigma, \quad x > M.$$

- *The denial of Definition 9 is*: A fixed number M is **not an upper bound** of a subset Σ of the set of real numbers, if

$$\sim (\forall x \in \Sigma, \quad x \le M),$$

that is if

$$\exists x \in \Sigma, \quad x > M . ♥$$

Example 12: In each case, give an upper bound of the set Σ.

1. $\Sigma = [0, 1[$
2. $\Sigma =]1, \infty[$
3. $\Sigma = \{x \in \mathbb{Q}^+ : x^2 < 2\}$

Solution

12.1. Clearly, every element of Σ is smaller than or equal to 1. Therefore, 1 is an upper bound of Σ; and by Axiom 17 of Chapter 9, so is every number $M > 1$.

12.2. It is easy to see that Σ is not bounded from above [by any real number]. All we need to convince ourselves of this is to notice that the denial of Definition 8 holds in this case. Indeed, for every real number M, we need to just choose $x = M + 1$, and we shall have $x > M$.

12.3. The set is bounded from above by 2. We can show this **by contradiction**. Indeed, if it were not bounded from above, then there would be in it at least one rational $r \in \Sigma$, such that $r > 2$. But then, this would imply that $r^2 > 4$, which clearly contradicts the fact that r, as a member of Σ, must satisfy $r^2 < 2$. ♥

Remark 13: It is well to notice at this point that if a set Σ has an upper bound M, then it has an infinite number of upper bounds. For, clearly, $M+1$, $M+2$, $M+3, \ldots$ would also all be upper bounds of Σ. ■

Hence, the following

Definition 14 [Least Upper Bound]: Let Σ be a non empty subset of $\mathbb{R}$. A real number M is called **the least upper bound** of Σ, if it satisfies both of these conditions:

1. M is an upper bound of Σ:

$$\forall x \in \Sigma, \quad x \leq M$$

2. M is the least of all upper bounds of Σ:

If K too is an upper bound of Σ, then $K \geq M$.

Remark 15: Note the use of the definite article "**the**" in the expression "the least upper bound." This presupposes that a set can have **at most one** least upper bound. That this is a fact will be proved in the next theorem. But, for the meantime, let us look at a couple of examples. ■

Example 16:
Show that 1 is the least upper bound of $\Sigma = [0, \ 1[$.

Solution:
We have already demonstrated that 1 is an upper bound of Σ. So now, we must show that any other upper bounded of Σ is necessarily greater than or equal to 1. We will use the methods of **contradiction** and **exhaution**. Hence, suppose K is another upper bound of Σ, but that $K < 1$. This gives us two cases:

Case 1. $K < 0$:
But this is a contradiction, since K, being an upper bound, must be greater than or equal to every element of Σ, including 0 for that matter.

Case 2. $0 \leq K < 1$:

Then $0 \leq K < \frac{K+1}{2} < 1$ which implies that there is at least one element of Σ, namely $\frac{K+1}{2}$, greater than the upper bound K. But this too is a contradiction with the fact that K is an upper bound of Σ. ♥

Example 17:
Show that the least upper bound of $A = \{x \in \mathbb{Q}^+ : x^2 < 2\}$, if it exists, cannot be a rational number.

Solution:
We will argue by **contradiction**. Hence, assume that A has a least upper bound M that belongs to $\mathbb{Q}$. Then

$$\forall x \in A, \quad x \leq M \in \mathbb{Q}. \tag{10}$$

Now, for the remainder of the proof, there are two cases:

Case 1. All items of $B = \{x \in \mathbb{Q}^+ : x^2 > 2\}$ are greater than or equal to M:
In this case, we would have

$$[\forall x \in A, \quad \forall y \in B, \quad x \leq M \leq y], \quad \textit{and} \quad M \in \mathbb{Q},$$

which contradicts the finding of Exercise 7.

Case 2. There is at least one element in B, say y_0, that is less than M:
In this case, since we have seen in Exercise 5.4 that every element of set A is smaller than every element of set B, we must therefore have

$$\forall x \in A, \quad x \leq y_0 < M.$$

This means that y_0 is a smaller upper bound of set A than M is. And clearly, this contradicts the assumption that $M \in \mathbb{Q}$ is the least upper bound of set A. ♥

We are now in a position to prove the uniqueness of the least upper bound.

Theorem 18 [Uniqueness of the Least Upper bound]: The least upper bound of any non empty subset Σ of $\mathbb{R}$, if it exists, is unique.

Proof:
We shall use the method of **contradiction**. Suppose Σ has more than one least upper bounds. This means Σ has at least two distinct least upper bounds, say M and K. Then, using Part 2 of Definition 14,

$$K \geq M , \quad \text{as } M \text{ is a least upper bound of } \Sigma \tag{11}$$

and

$$M \geq K , \quad \text{as } K \text{ is a least upper bound of } \Sigma \tag{12}$$

But then, (11) and (12) imply that $K = M$[13], which clearly contradicts the fact that M and K are distinct numbers. Thus, Σ cannot have more than one least upper bound. ♣

Our next theorem, the **Least Upper Bound Principle**, is very important. It is the culmination of this section. Some writers use it as an axiom in lieu of the Axiom of Continuity. In fact, it is not difficult to show that this principle is equivalent to the Axiom of Continuity. The least upper bound principle is used to establish many important properties of the real numbers.

Theorem 19 [The Least Upper Bound Principle]: Every nonempty subset Σ of $\mathbb{R}$ that is bounded from above [by some real number M] has a least upper bound in $\mathbb{R}$ [which may or may not equal M .]

Proof:
Let Σ be a nonempty subset of $\mathbb{R}$ that is bounded from above by some number $M \in \mathbb{R}$. Now, denote by S the set of all upper bounds of Σ.

[13] Refer to Exercise 38.3 of Chapter 9.

[Note that S is not empty. Indeed by definition, S contains the number M .] Thus, we have

$$\forall x \in \Sigma, \quad \forall s \in S, \quad x \leq s$$

Hence, by the axiom of continuity, we deduce that there is a fixed real number s_0 such that

$$\forall x \in \Sigma, \quad \forall s \in S, \quad x \leq s_0 \leq s$$

This means that s_0 is not only an upper bound of the set Σ but also the least among all the upper bounds $s \in S$. Thus, the least upper bound of Σ exists and is s_0 . ♣

Example 20:
Show that $A = \{x \in \mathbb{Q}^+ : x^2 < 2\}$ has a least upper bound.

Solution:
We saw in Example 12.3 that set A is bounded from above. Thus, by the **Least Upper Bound Principle**, it has a least upper bound. ♥

11.5. GREATEST LOWER BOUND PRINCIPLE

This section is very similar to the preceding section. It does exactly for the lower bounds what the preceding section did for upper bounds.

Definition 21 [Lower Bounds]: A nonempty subset Σ of $\mathbb{R}$ is said to be **bounded from below**, if

$$\exists m \in \mathbb{R}, \quad \forall x \in \Sigma, \quad m \leq x. \tag{13}$$

And the fixed real number m is called a **lower bound** of Σ [since every element of Σ is greater than or equal to m .]

Example 22: In each case, give a lower bound of the set Σ

1. $\Sigma =]0, 1]$

2. $\Sigma =]-\infty, \ 0[$

3. $\Sigma = \{x \in \mathbb{Q}^+ : \ x^2 > 2\}$

Solution

This example is pretty similar to Example 12. So, we leave it to the reader to supply all the details of the solution. ♥

Remark 23: Notice that if a set Σ has a lower bound m, then it has an infinite number of lower bounds. For, indeed, the numbers $m-1$, $m-2$, $m-3$,... would also all be lower bounds of Σ. ■

The foregoing remark leads us to the following

Definition 24 [Greatest Lower bound]: Let Σ be a nonempty subset of $\mathbb{R}$. A fixed real number m is called **the greatest lower bound** of Σ, if it satisfies both of these conditions:

1. m is a lower bound of Σ:

$$\forall x \in \Sigma, \quad m \leq x$$

2. m is the greatest of all lower bounds of Σ:

If k too is a lower bound of Σ, then $k \leq m$.

Example 25:

Show that 0 is the greatest lower bound of $\Sigma =]0, \ 1]$.

Solution:

This example is pretty similar to Example 16. So, we leave it to the reader to supply all the details. ♥

Example 26:

Show that the greatest lower bound of $B = \{x \in \mathbb{Q}^+ : \ x^2 > 2\}$ if it exists, cannot be a rational number.

Solution:
This example is pretty similar to Example 17. So, we leave it to the reader to supply all the details. ♥

The following theorem proves the uniqueness of greatest lower bounds.

Theorem 27 [Uniqueness of the Greatest Lower Bound]: The greatest lower bound of any nonempty subset Σ of $\mathbb{R}$, if it exists, is unique.

Proof:

We shall use the method of **Proof by Contradiction**. Suppose Σ has more than one greatest lower bound. This means Σ has at least two distinct greatest lower bounds, say m and k. Then, using part 2 of Definition 24, we get

$$k \leq m, \quad \text{as } m \text{ is a greatest lower bound of } \Sigma \tag{14}$$

and

$$m \leq k, \quad \text{as } k \text{ is a greatest lower bound of } \Sigma \tag{15}$$

But then, (14) and (15) imply that $k = m$, which clearly contradicts the fact that m and k are distinct numbers. Hence, Σ does have a unique greatest lower bound. ♣

We now introduce another important consequence of the Axiom of Continuity. It is called the **Greatest Lower Bound Principle** and is pretty similar to the Least Upper Bound Principle.

Theorem 28 [Greatest Lower Bound Principle]: Every nonempty subset Σ of $\mathbb{R}$ that is bounded from below [by some real number m] has a greatest lower bound in $\mathbb{R}$ [which may or may not equal m.]

Proof:
Let Σ be a nonempty subset of $\mathbb{R}$ that is bounded from below by some number $m \in \mathbb{R}$. Denote by I the set of all lower bounds of Σ. [Note that I is not empty, since I contains the number m.] Thus, we have

$$\forall x \in \Sigma, \quad \forall i \in I, \quad i \leq x$$

Hence, by the Axiom of Continuity, we know that there is a fixed real number i_0 such that

$$\forall x \in \Sigma, \quad \forall i \in I, \quad i \leq i_0 \leq x.$$

This means that i_0 is not only a lower bound of the set Σ but also the greatest among all the lower bounds $i \in I$. Thus, the greatest lower bound of Σ exists and is i_0. ♣

Example 29:
Show that $B = \{x \in \mathbb{Q}^+ : \ x^2 > 2\}$ has a greatest lower bound.

Solution:
Clearly, the set B is bounded from below by 0. Thus, by the Greatest Lower Bound Principle, B has a greatest lower bound. ♣

11.6. ARCHIMEDEAN PRINCIPLE

In this section we will prove another important property of real numbers. This property is called the **Archimedean principle.** We shall also explore the consequences of this principle. Basically, what this property says is that given any real number x and any positive real number a, it is possible to add a to itself repeatedly till one exceeds x.

Theorem 30 [First Archimedean Principle]: let us be given two fixed numbers x and a in $\mathbb{R}$. If $a > 0$, then there is a **unique** integer k such that :

$$ka \leq x < (k+1)a.$$

Proof:
We will argue **by contradiction**.

Step one: Assume $ka \leq x$ for all $k \in \mathbb{Z}$. This means that the set

$$A = \{ ka \ : \ k \in \mathbb{Z}\}$$

is bounded from above by x. Thus, by the Least Upper Bound Principle, A has a least upper bound, say M. Now, since $M - a$ is less than M , $M - \alpha$ is not an upper bound of A . [For, if it were, it would then be a lesser upper bound than the least upper bound M , which is impossible]. Hence, there is at least one $p \in \mathbb{Z}$ such that

$$pa > M - a \ ;$$

that is

$$(p+1)\,a \ > M \, . \tag{16}$$

But this clearly contradicts the earlier statement that M is an upper bound of set A. From this contradiction we conclude that our assumption at the beginning is false. Thus, there is at least one integer k_0 such that

$$x \ < \ k_0 \, a \, . \tag{17}$$

Step two: Now, Assume similarly that $x \leq na$ for all $n \in \mathbb{Z}$. This too means that the set

$$A = \{ na \ : \ n \in \mathbb{Z}\}$$

is bounded from below by x. Thus, by the Greatest Lower Bound Principle, A has a greatest lower bound, say m. Now, since $m + a$ is greater than m, $m + a$ is not a lower bound of A. [For, if it were, it would then be a greater lower bound than the greatest lower bound m, which is impossible]. Hence, there is at least one $q \in \mathbb{Z}$ such that

$$qa < m + a \ ;$$

that is

$$(q-1)a \ < \ m \ . \tag{18}$$

But this clearly contradicts the earlier statement that m is a lower bound of set A. From this contradiction we conclude that our beginning assumption is false. Thus, there is at least one integer n_0 such that

$$n_0\, a < x\,. \tag{19}$$

Step three: Now combining (18) and (19), we get

$$n_0\, a < x < k_0\, a\,.$$

Therefore, by checking each of these intervals, $[n_0 a,\ (n_0+1)a[$, $[(n_0+1)a,\ (n_0+2)a[$, ..., and $[(k_0-1)a,\ k_0 a[$, we finally find one and only one among them, say $[ka,\ (k+1)a[$, which contains the number x. In other words,

$$ka \le x < (k+1)a\,. \clubsuit$$

Exercise 31:

1. Let α be any real number. Show that there is an integer n_0 such that: $n_0 > \alpha$.
2. Let $\varepsilon > 0$ be any positive real number. Show that there is an integer n_0 such that:

$$0 < \frac{1}{n_0} < \varepsilon\,.$$

Solution

31.1. By applying Theorem 30, with $a = 1$ and $x = \alpha$, we have:

$$\exists k \in \mathbb{Z}, \quad k \le \alpha < k+1$$

Thus, by setting $n_0 = k+1$, we arrive at the desired result: $n_0 > \alpha$.

31.2. If $\varepsilon > 0$, then $\varepsilon \neq 0$ and so is $\frac{1}{\varepsilon}$ a real positive number. Therefore, by the last result just established above, we have

$$\exists n_0 \in \mathbb{Z}, \quad n_0 > \frac{1}{\varepsilon}\,,$$

that is

$$\exists n_0 \in \mathbb{Z}, \quad 0 < \frac{1}{n_0} < \varepsilon\,. \heartsuit$$

In reality, we have not just one Archimedean principle, but two: (1) A "multiplicative" version presented as Theorem 30, and an "exponential" one given below.

Theorem 32 [Second Archimedean Principle]: let x and a be in $\mathbb{R}$. Suppose $x > 0$, and $a > 1$ then there is a unique integer k such that:

$$a^k \leq x < a^{k+1}.$$

Proof:

Left as an exercise. [See Problem 6 at the end of this Chapter]. ♣

Exercise 33:

1. Suppose $\alpha > 0$ and $M > 1$. Show that there is an integer n_0 such that:
$$M^{n_0} > \alpha .$$
2. Suppose $\varepsilon > 0$ and $0 < b < 1$. Show that there is an integer n_0 such that:
$$0 < b^{n_0} < \varepsilon .$$

Solution

33.1. By applying Theorem 32, we find that there is an integer k such that

$$M^k \leq \alpha < M^{k+1}.$$

Therefore, setting $n_0 = k + 1$, we get the desired result: $M^{n_0} > \alpha.$

33.2. Since $0 < b < 1$, we have $\frac{1}{b} > 1$. Therefore, by applying Theorem 32, we get

$$\left[\frac{1}{b}\right]^k \leq \frac{1}{\varepsilon} < \left[\frac{1}{b}\right]^{k+1}$$

for some integer k. Thus

$$0 < b^{k+1} < \varepsilon \ .$$

Therefore, by setting $n_0 = k + 1$, we get what we wanted: $0 < b^{n_0} < \varepsilon$. ♥

11.7. SOME DENSE SUBSETS OF THE SET OF REALS NUMBERS

In this last section we will look at an interesting idea called **density**. Density in mathematics is not quite the same as density in geography, but the two concepts are somehow intuitively similar. In mathematics a subset S of $\mathbb{R}$ is considered dense in $\mathbb{R}$, if for any given $x \in \mathbb{R}$, there is an element s of S that is as "close" to x as we desire. And similarly, in geography, we are inclined to say that a land is densely populated, if the likelihood of finding people living close to any given location on that land is pretty high.

The following theorem shows that the set $\mathbb{Q}$ of rational numbers is *dense* in the set $\mathbb{R}$ of real numbers.

Theorem 34 [$\mathbb{Q}$ is dense in $\mathbb{R}$]. Every open interval $]a, b[$, where $a < b$ and a and b are in $\mathbb{R}$, contains at least one rational number.

Proof:
To prove this theorem, we will use the first **Archimedean principle.**

Step one. Let n be an integer such that

$$0 < \frac{1}{n} < b - a \tag{20}$$

[The existence of such an integer n was shown in Exercise 31.2.]

Step two. By using the first Archimedean Principle, we see that there is an integer k such that

$$k\frac{1}{n} \leq a < (k+1)\frac{1}{n} \tag{21}$$

Now, adding $1/n$ to and subtracting a from both sides of the first inequality in (21), we have

$$\frac{k+1}{n} - a \leq \frac{1}{n} \tag{22}$$

Hence, combing (20) and (22) we arrive at

$$\frac{k+1}{n} - a < b - a\,. \tag{23}$$

Thus,

$$\frac{k+1}{n} < b \tag{24}$$

Step three. Finally, using (24) and the second inequality in (21), we also have:

$$a < \frac{k+1}{n} < b\,.$$

In other words, we have **constructed** a rational number, namely $(k+1)/n$, which belongs to the open interval $]a,\ b[$. Hence the proof is complete. ♣

Remark 35: In reality, there is an infinite number of rational numbers in any open interval $]a,\ b[$. To see this, it suffices to reapply the preceding theorem to the interval $\left]\frac{k+1}{n},\ b\right[$. We obtain a new rational number r such that.

$$\frac{k+1}{n} < r < b$$

This process is then repeated for the interval $]r,\ b[$, and indefinitely afterwards . ■

We have just proved that the set $\mathbb{Q}$ of rationals is dense in the set $\mathbb{R}$ of real numbers. So, to close this chapter, it will be interesting to also demonstrate that the set $\mathbb{I}$ of irrationals is dense in $\mathbb{R}$. Our next theorem does just that.

Theorem 37 [$\mathbb{I}$ is dense in $\mathbb{R}$]. Every open interval $]a,\ b[$, where $a < b$ and a and b are in $\mathbb{R}$, contains at least one irrational number.

Proof by construction:

Step one. Let us choose in $]a,\ b[$ any two distinct rational numbers, say r and s with $r < s$. [The existence of such rational numbers is guaranteed by Theorem 34 and Remark 35.]

Step two. Now construct a number u as follows

$$u = r + \frac{\sqrt{2}}{n}(s - r) \tag{25}$$

where n is any fixed natural number greater than or equal to 2 .

Next, it suffices to show that u , as constructed, is an irrational number in the interval $]a,\ b[$.

Step three. Clearly,

$$r < u. \tag{26}$$

Also, since $n \geq 2 > \sqrt{2}$, we have

$$\frac{\sqrt{2}}{n} < 1 . \tag{27}$$

Now, multiply both sides of (27) by $s - r$ and add r to the products; we obtain

$$u = r + \frac{\sqrt{2}}{n}(s - r) < r + (s - r) = s ;$$

that is

$$u < s. \tag{28}$$

Thus, combining (26) and (28) we get $r < u < s$ which shows that $u \in]a,\ b[$.

Step four. Therefore, the only thing that remains to be shown is that u is irrational. Let us prove this by **contradiction**. Suppose u is not irrational, then u is rational. But, this would mean, by solving equation (25) for $\sqrt{2}$, that

$$\sqrt{2} = \frac{n(u-r)}{s-r} \in \mathbb{Q},$$

since u, s, r and n are all rationals. But this is impossible, for $\sqrt{2}$ is not rational. Thus, the proof is complete. ♣

PROBLEMS

1. Prove that the *least upper bound* of set $A = \{x \in \mathbb{Q}^+ :\ x^2 < 2\}$ and the *greatest lower bound* of set $B = \{x \in \mathbb{Q}^+ :\ x^2 > 2\}$ are the same real number.

2. **Existence of the real number $\sqrt{2}$:** Let us denote by α the *least upper bound* of set $A = \{x \in \mathbb{Q}^+ :\ x^2 < 2\}$. Show that $\alpha^2 = 2$.

3. Prove that the *Least Upper bound Principle*, the *Greatest Lower Bound Principle*, and *the Axiom of Continuity* are equivalent. In other words, you may want to show these propositions:

   3.1. The least upper bound principle $\Rightarrow$ The greatest lower bound principle.

   3.2. The greatest lower bound principle $\Rightarrow$ The Axiom of Continuity

and

3.3. The Axiom of Continuity $\Rightarrow$ The least upper bound principle. [You need not prove this last implication; for it is Theorem 19 of this Chapter.]

4. * Let S be a nonempty set of real numbers bounded from above and α any positive real number. Prove that if M is the least upper bound of S, then there is at least one number s in S such that:

$$M - \alpha < s \leq M.$$

5. Let S be a nonempty set of real numbers bounded from below and α any positive real number. Prove that if m is the greastest lower bound of S, then there exists at least one number s in S such that:

$$m \leq s < m + \alpha.$$

6. Prove Theorem 32 of this chapter.

7. Prove that $3 > 2$. [Hint: You may use the definition: $3 = 2 + 1$.]

8. More generally, show that: For every $n \in \mathbb{N}$, $n + 1 > n$.

"There is a tradition of opposition between adherents of induction and of deduction. In my view it would be just as sensible for the two ends of a worm to quarrel."
Alfred North Whitehead.

CHAPTER 12

METHODS OF PROOF BY MATHEMATICAL INDUCTION

In Chapter 3, we discussed a number of proof methods. In this chapter, we will study yet another powerful and frequently used method of proof known as **mathematical induction**.

12.1. THE SYSTEM OF NATURAL NUMBERS

In the preceding chapter, we defined the set $\mathbb{N}$ of natural numbers as:

$$\mathbb{N} = \{\ 1,\quad 1+1,\quad (1+1)+1,\quad ((1+1)+1)+1,\ ...\ \}$$

Thus, 1 is the smallest element of $\mathbb{N}$.

We now state a theorem whose role in this chapter will be crucial.

Theorem 1 [The Least Number Principle[14]]: Every nonempty subset S of $\mathbb{N}$ contains an element that is smaller than all the other elements of S.

[14] This principle is sometimes called *The Well-Ordering Principle*.

To prove this theorem, we will make use of the greatest lower bound principle of the preceding chapter.

Proof:
Let S be a nonempty subset of $\mathbb{N}$. We know that S is bounded from below by 1, since 1 is the smallest natural number. Therefore, by the greatest lower bound principle, S has a greatest lower bound, say l. If we can show that l belongs to S, this will establish the theorem. To do so, let us argue by **contradiction**.

So, let us assume that $l \notin S$. Then, since $l < l+1$ and l is the greatest lower bound, we deduce that $l+1$ is not a lower bound of S. Therefore, there is at least one element x of S such that:

$$l \leq x < l+1 \qquad (1)$$

But, by assumption, l is not in S, so then l cannot equal x. Therefore, the first inequality in (1) must be strict. That is

$$l < x < l+1 .$$

Similarly, x, being greater than the greatest lower bound l, cannot be a lower bound of the set S. Thus, again there is at least one element, say y, of S such that

$$l \leq y < x < l+1 . \qquad (2)$$

And again, since l is not in S, the first inequality in (2) must be strict. Therefore

$$l < y < x < l+1 ,$$

which implies that

$$0 < x - y < 1, \quad \text{where} \quad x - y \in \mathbb{N}.$$

Thus, we have arrived at the conclusion that the natural number $x-y$ is smaller than 1, which is clearly a contradiction. For, 1 is the smallest natural number. Hence, the proof is complete. ♣

12.2. THE FIRST PRINCIPLE OF INDUCTION

We are now in the position to give our first principle on which proof by mathematical induction is based.

> **Theorem 2 [First Principle of Mathematical Induction]:** Let $P(n)$ be an open sentence about the variable $n \in \mathbb{N}$.
> If
> 1. $P(1)$ is true
>
> and
>
> 2. $P(k) \Rightarrow P(k+1)$,
>
> then
>
> $(\forall n \in \mathbb{N}, \quad P(n))$ is true.

Remark 3: To get the general idea of this principle, it may be helpful to think of a railway train with an infinite number of cars, and to interpret the hypotheses of Theorem 2 as follows: **Hypothesis 1** ascertains that the first car is moved [by an engine.] Then **Hypothesis 2** guarantees that if any car whatsoever, say the k th one, is moved then so is the next $(k+1)$ th car. Under those hypothesis, one would want to know what would then happen to the overall train? Of course, it is not difficult to see that, together, these two hypothesis are enough to ensure that all of the cars will move simultaneously. ■

Proof:

We shall argue by **contradiction**. Thus, let us assume that there is at least one natural number n_0 such that $P(n_0)$ is false. Note that n_0 must be different from 1 ; for indeed, $P(1)$ is true. Now, let us denote by S the set of all natural numbers n for which the open sentence $P(n)$ is false:

$$S = \{ n \in \mathbb{N} : \quad P(n) \ is \ false \}$$

Clearly, S is nonempty since, by the assumption that $P(n_0)$ is false, $n_0 \in S$. Now, S being a nonempty subset of $\mathbb{N}$, we know from the

Least-Number Principle that S has a smallest element, say s_0 such that

$$P(s_0) \text{ is false.} \tag{3}$$

On the other hand, s_0 being the smallest element of S, $s_0 - 1$ does not belong to S. Thus,

$$P(s_0 - 1) \text{ is true.} \tag{4}$$

But then, using the hypothesis $P(k) \Rightarrow P(k+1)$, we conclude from (4) that

$$P(s_0) \text{ is true,}$$

which contradicts (3). Hence the proof is complete. ♣

Remark 4: Thus, a proof by mathematical induction may consist of three steps:

Step 1 [or *The Initial step*] is the verification that the open sentence $P(n)$ is true for $n = 1$.

Step 2 [or *The Induction hypothesis*] is the assumption that the open sentence $P(n)$ is be true for $n = k$, where $k \in \mathbb{N}$.

Step 3 [or *The Inductive step*] is the proof that $P(n)$ is true for $n = k+1$, by virtue of the assumption made in Step 2.

Example 5:
Prove the following formula by mathematical induction:
For all natural numbers n,

$$1 + 2 + 3 + \ldots + n = \frac{n(n+1)}{2}.$$

Solution:
Let us denote by $P(n)$ the open sentence:

$$1 + 2 + 3 + \ldots + n = \frac{n(n+1)}{2}.$$

Initial step. From the above we see that $P(1)$ is the proposition

$$1 = \frac{1(1+1)}{2}$$

Thus, $P(1)$ is true.

Induction hypothesis. Assume $P(k)$ is true for some $k \in \mathbb{N}$. That is, assume

$$1 + 2 + 3 + \ldots + k = \frac{k(k+1)}{2}$$

is true for some $k \in \mathbb{N}$.

Inductive step. Now, let us deduce from the *Induction hypothesis* that $P(k+1)$ too holds true:

$$\begin{aligned} 1 + 2 + 3 + \ldots + k + (k+1) &= [1 + 2 + 3 + \ldots + k] + (k+1) \\ &= \frac{k(k+1)}{2} + (k+1), \text{ by the Induction hypothesis.} \\ &= \frac{k(k+1)}{2} + \frac{2(k+1)}{2} \\ &= \frac{(k+1)[(k+1)+1]}{2}. \end{aligned}$$

Thus, assuming that $P(k)$ is true does indeed make $P(k+1)$ too true.

Therefore, by Theorem 2, we have shown that:

$$1 + 2 + 3 + \ldots + n = \frac{n(n+1)}{2}, \text{ for all } n \in \mathbb{N}. ♥$$

Example 6:
Prove the following by mathematical induction:

> For all natural numbers n, $5 \cdot 2^{3n-2} + 3^{3n-1}$ is divisible by 19.

Solution:
Let us denote by $P(n)$ the open sentence:

$$5 \cdot 2^{3n-2} + 3^{3n-1} \text{ is divisible by } 19.$$

Initial step. From the above we see that $P(1)$ is the proposition:

$$5 \cdot 2^{1} + 3^{2} \text{ is divisible by } 19.$$

Thus, $P(1)$ is true.

Induction hypothesis. Assume $P(k)$ is true for some $k \in \mathbb{N}$. That is, assume

$$5 \cdot 2^{3k-2} + 3^{3k-1} \text{ is divisible by } 19$$

is true for some $k \in \mathbb{N}$.

Inductive step. Now, let us deduce from the *Induction hypothesis* that $P(k+1)$ too is true:

$$\begin{aligned} 5 \cdot 2^{3(k+1)-2} + 3^{3(k+1)-1} &= 5 \cdot 2^{3k-2}2^{3} + 3^{3k-1}3^{3} \\ &= 8 \cdot 5 \cdot 2^{3k-2} + 27 \cdot 3^{3k-1} \\ &= 8 \cdot 5 \cdot 2^{3k-2} + 8 \cdot 3^{3k-1} \\ &\quad + 27 \cdot 3^{3k-1} - 8 \cdot 3^{3k-1} \\ &= 8 \cdot \left[5 \cdot 2^{3k-2} + 3^{3k-1}\right] + 19 \cdot 3^{3k-1} \end{aligned}$$

.

Thus, assuming that $P(k)$ is true makes $P(k+1)$ too true.

Therefore, by Theorem 2, we have shown that:

$5 \cdot 2^{3n-2} + 3^{3n-1}$ is divisible by 19, for all $n \in \mathbb{N}$. ♥

12.3. THE SECOND PRINCIPLE OF INDUCTION

Our next theorem, the **Second Principle of Mathematical Induction**, is a simple but useful generalization of the first principle.

Theorem 7 [Second Principle of Mathematical Induction]: Let n_0 be a fixed natural number, and $P(n)$ an open sentence about all $n \in \mathbb{N}$ such that $n \geq n_0$.
If

1. $P(n_0)$ is true

and

2. $P(k) \Rightarrow P(k+1)$,

then

$(\forall n \geq n_0, \quad P(n))$ is true.

Remark 8: Note that in the case where $n_0 = 1$, Theorem 7 is simply Theorem 2. ■

Proof [of Theorem 7]:

Step one. Let us introduce a new variable m defined by

$$m = n - n_0 + 1.$$

Then, it is easy to see that m takes the values 1, 2, 3, . . . , as the variable n assumes the values n_0, $n_0 + 1$, $n_0 + 2, \ldots$; and vice versa.

Step two. Now, denote by $S(m)$ the open sentence $P(m + n_0 - 1)$. Then we have the equivalences:

$$\begin{cases} P(n_0) \Leftrightarrow S(1) \\ \textit{and} \\ [P(n) \Rightarrow P(n+1), \quad n \geq n_0] \Leftrightarrow [S(m) \Rightarrow S(m+1), \quad m \geq 1]. \end{cases}$$

Step three. Hence, from the assumption that

$$\begin{cases} P(n_0) \text{ is true} \\ \text{and} \\ P(n) \Rightarrow P(n+1), \quad n \geq n_0 \end{cases}$$

we deduce that

$$\begin{cases} S(1) \text{ is true} \\ \text{and} \\ S(m) \Rightarrow S(m+1), \quad m \geq 1. \end{cases}$$

Thus, by virtue of Theorem 2, we have shown the following to be true:

$$\forall m \in \mathbb{N}, \ S(m),$$

which is the same as proving

$$\forall n \geq n_0, \ P(n). \ ♣$$

Example 9:
Prove the following by mathematical induction:

For all natural numbers $n \geq 4$, $\quad 2^n < n!$

Solution:
Let us denote by $P(n)$ the open sentence:

$$2^n < n!$$

It is easy to verify that the smallest natural number for which $P(n)$ is true is $n = 4$. Indeed,

$P(1) \Leftrightarrow 2^1 < 1!$, which is false

$P(2) \Leftrightarrow 2^2 < 2!$, which is false

$P(3) \Leftrightarrow 2^3 < 3!$, which is false

$P(4) \Leftrightarrow 2^4 < 4!$, which is **true.**

Initial step. From the above we saw that $P(4)$ is a true proposition.

Induction hypothesis. Assume $P(k)$ is true for some $k \geq 4$. That is, assume

$$2^k < k!$$

is true for some $k \geq 4$.

Inductive step. Now, let us deduce from the *Induction hypothesis* that $P(k+1)$ too is true.

$$\begin{aligned} 2^{(k+1)} &= 2 \cdot 2^k \\ &< 2(k!) \text{ , by the Induction hypothesis} \\ &< (k+1) \cdot (k!) \text{ , since } 2 < k+1 \text{ for all } k \geq 4 \\ &= (k+1)! \end{aligned}$$

Thus, assuming that $P(k)$ is true implies $P(k+1)$ too is true. Therefore, by Theorem 7, we have shown that:

$$2^n < n!, \quad \text{for all } n \geq 4 \text{ . } \heartsuit$$

Example 10:
Prove the following by mathematical induction.

$$\text{For all } n \in \mathbb{N} \qquad 2^n > n.$$

Solution:
Let us denote by $P(n)$ the open sentence:

$$2^n > n.$$

Initial step. From the above we see that $P(1)$ is the proposition:

$$2^1 > 1$$

Thus, $P(1)$ is true.

Induction hypothesis. Assume $P(k)$ is true for some $k \in \mathbb{N}$. That is, assume

$$2^k > k$$

is true for some $k \in \mathbb{N}$.

Inductive step. Now, let us deduce from the **Inductive hypothesis** that $P(k+1)$ too is true.

$$\begin{aligned} 2^{(k+1)} &= 2 \cdot 2^k \\ &> 2k \ , \quad \text{by the } \textit{Induction hypothesis} \\ &= k + k \\ &\geq k+1, \quad \text{since } k \geq 1 \text{ for all } k \geq 4 \end{aligned}$$

Thus, assuming that $P(k)$ is true makes $P(k+1)$ too true.

Therefore, by Theorem 7, we have shown that: $2^n > n$, for all $n \in \mathbb{N}$. ♥

12.4. THE THIRD PRINCIPLE OF INDUCTION

Other problems require a yet another generalization of the First Principle of Mathematical induction. We present this generalization in our next theorem.

Theorem 11 [Third Principle of Mathematical Induction]: Let n_0 be a fixed natural number, and $P(n)$ an open sentence about all $n \geq n_0$.

If

1. $P(n_0)$ is true

and

2. $P(n_0)$, $P(n_0+1)$, $P(n_0+2)$,... and $P(k) \Rightarrow P(k+1)$,

then

($\forall n \in \mathbb{N}, \quad P(n)$) is true.

Proof:
The proof of this third principle of mathematical induction is pretty similar to the proof of the first principle. We therefore leave it to the reader to supply all the details. ♣

Example 12:
Prove the following by mathematical induction:

For all $n \in \mathbb{N}$,
$T_n(x) = \cos[n \cdot \text{Arcos}(x)]$, $\qquad x \in [-1,\ 1]$
is a polynomial of degree $n \in \mathbb{N}$.

Solution
Let us denote by $P(n)$ the open sentence:
$T_n(x) = \cos[n \cdot \text{Arcos}(x)]$, $\qquad x \in [-1,\ 1]$
is a polynomial of degree $n \in \mathbb{N}$.

Initial step. From the above we see that $P(1)$ is the proposition:

$$T_1(x) = \cos[\text{Arcos}(x)] = x, \qquad x \in [-1,\ 1] \qquad (5)$$

is a polynomial of degree 1 .

Thus, $P(1)$ is true.

Induction hypothesis. Assume $P(1)$, $P(2)$, $P(3)$,... and $P(k)$ are all true for some $k \in \mathbb{N}$. That is, assume

$$T_j(x) = \cos[j \cdot \text{Arcos}(x)], \qquad x \in [-1,\ 1].$$

is a polynomial of degree $j \in \mathbb{N}$, for every $j \leq k$, where k is some natural number.

Inductive step. We will deduce from the Induction hypothesis that $P(k+1)$ too is true. Notice that

$$T_{k+1}(x) \quad = \quad \cos[(k+1) \cdot \text{Arcos}(x)]$$

$$
\begin{aligned}
&= \cos[k \cdot \text{Arcos}(x) + \text{Arcos}(x)] \\
&= \cos[k \cdot \text{Arcos}(x)] \cos[\text{Arcos}(x)] \\
&\quad - \sin[k \cdot \text{Arcos}(x)] \sin[\text{Arcos}(x)] \\
&= T_k(x) \cdot T_1(x) - \sin[k \cdot \text{Arcos}(x)] \sin[\text{Arcos}(x)] \\
&= T_k(x) \cdot T_1(x) - \frac{1}{2}[\cos[(k-1) \cdot \text{Ar}\cos(x)]] \\
&\quad + \frac{1}{2}[\cos[(k+1) \cdot \text{Ar}\cos(x)]] \\
&= T_k(x) \cdot T_1(x) - \frac{1}{2}[T_{k-1}(x) - T_{k+1}(x)]
\end{aligned}
$$

That is

$$T_{k+1}(x) = T_k(x) \cdot T_1(x) - \frac{1}{2}[T_{k-1}(x) - T_{k+1}(x)] \qquad (6)$$

Therefore, by solving (6) for $T_{k+1}(x)$ we get:

$$T_{k+1}(x) = 2T_1(x) \cdot T_k(x) - T_{k-1}(x),$$

which, by using (5), can be rewritten as

$$T_{k+1}(x) = 2xT_k(x) - T_{k-1}(x) . \qquad (7)$$

But, from the *Induction hypothesis*, we know that $T_k(x)$ is a polynomial of degree k and $T_{k-1}(x)$ too a polynomial of degree $k-1$. Thus, we deduce from (7) that $T_{k+1}(x)$ is a polynomial of degree $k+1$.

Therefore, by Theorem 11,

$$\forall n \in \mathbb{N}, \quad T_n(x) = \cos[n \cdot \text{Arcos}(x)], \qquad x \in [-1, 1]$$

is a polynomial of degree $n \in \mathbb{N}$. This completes the proof. ♥

PROBLEMS

1. Prove that any set S of integers bounded from below has a least number.

2. Prove *The Division Algorithm*: If a and b are natural numbers and $a > b$, then there exist unique integers q and r such that

$$a = qb + r \text{ and } 0 < q \text{ and } 0 \leq r < q .$$

3. Use mathematical induction to prove each of the following relations:

3.1. $n^5 - n$ *is divisible by* 5, $n \in \mathbb{N}$.

3.2. $1^3 + 2^3 + 3^3 + \ldots + n^3 = \left[\frac{n(n+1)}{2}\right]^2$, $n \in \mathbb{N}$.

3.3. $\frac{1}{n+1} + \frac{1}{n+2} + \frac{1}{n+3} + \ldots + \frac{1}{3n+1} > 1$, $n \in \mathbb{N}$.

3.4. $n(2n^2 - 3n + 1)$ *is divisible by* 6, $n \in \mathbb{N}$.

3.5. $n^7 - n$ *is divisible by* 7, $n \in \mathbb{N}$.

3.6. $2 + 2^2 + 3^3 + \ldots + 2^n = 2^{n+1} - 2$, $n \in \mathbb{N}$.

3.7. $\frac{x_1 + x_2 + \ldots + x_n}{n} \geq \sqrt{x_1 x_2 \ldots x_n}$, where $x_1, x_2, \ldots, x_n$ are any n positive real numbers, and $n \in \mathbb{N}$.

3.8. $2^n > n^2$, $n \in \mathbb{N}$ *and* $n \geq 5$.

3.9. Let there be a set of $n \geq 2$ lines in the plane no two of which are parallel, and no three of which pass through a same point. Then the number of points of intersection is $(n^2 - n)/2$.

3.10. We define the symbol C_n^r to be equal to:

$$C_n^r = \frac{n!}{r!(n-r)!}, \quad r \in \mathbb{N}, \ n \in \mathbb{N} \text{ and } r < n.$$

Then for $n \geq 2$,

$$C_{n+1}^{r} = C_{n}^{r} + C_{n}^{r-1}.$$

3.11. We define the function $f : \mathbb{Z} \to \mathbb{R}$ by:

$$f(1) = 3, \quad f(2) = \frac{3}{2}, \quad \textit{and} \quad f(n) = \frac{f(n-1) + f(n-2)}{2} \quad \textit{if } n > 2.$$

Then show that

$$f(n) = 2 + \left(-\tfrac{1}{2}\right)^{n-1}, \quad n \in \mathbb{N} \textit{ and } n \geq 2.$$

3.12. $2^{n} > 2n + 1$, $n \in \mathbb{N}$ *and* $n \geq 3$.

4. Let a and b be two natural numbers.

1. a is said to be a **divisor** of b, if there is some natural number k such that $b = ka$
2. a and b are said to be **co-primes**, if the only natural number divisor they have in common is 1.

Prove that if a and b are coprimes, then there are two integers x and y such that:

$$ax + by = 1.$$

5. Show that if a divides bc, and a and b are co-primes; then a divides c.

6. * Assume we are given:

1. a function $f : \mathbb{R} \to \mathbb{R}$ such that for every x and y in $\mathbb{R}$,

$$| f(x) - f(y) | \leq r | x - y |, \qquad 0 < r < 1.$$

2. and an infinite list of real numbers $(s_1, s_2, s_3, \ldots)$ defined by:

$$s_1 \in \mathbb{R}, \quad \textit{and} \quad s_n = f(s_{n-1}) \quad \textit{for all } n \geq 2$$

Then show by mathematical induction that

$$| s_{n+2} - s_{n+1} | \leq r^n | s_2 - s_1 |, \quad n \geq 1.$$

7. Prove Theorem 11 of this chapter.

"Mathematics is like checkers in being suitable for the young: not too difficult, amusing, and without peril to the state."
Plato.

CHAPTER 13

RELATIONS

13.1. RELATIONS

In our everyday conversations, we use phrases that describe how one person or object is related to another person or object. For example, in the sentence

Adam **is the husband of** Eve,

the relationship between Adam and Eve is described by the phrase "...is the husband of..." This phrase is the **link** between the first and second elements of the ordered pair:

$$(Adam,\ Eve).$$

Clearly, there are other ordered pairs which belong to this same relation. Indeed, any ordered pair

$$(M,\ W),$$

whatsoever in which M is a man, W a woman, and M is married to W does belong to this relation.

Definition 1: Let X and Y be any two sets. In mathematics, a **relation from X into Y** is simply any set R of ordered pairs (x, y) such that $x \in X$ and $y \in Y$.

Remark 2: In other words, a relation R from a set X into a set Y is merely a subset of the Cartesian product $X \times Y$. ■

Definition 3: Let R be a relation from a set X into a set Y , and (a, b) an ordered pair. We shall say that **a is related to b through R**, and write $a R b$, if $(a,\ b) \in R$.

Example 4: Suppose

$A = \{\ Pete,\ Danny,\ Ann,\ \ Nicole\ \}$ and $B = \{\ 7,\ 3,\ \ 5\ \}$.

Tell which ones of these sets are relations from A into B .

1. $R = \{\ (Pete, 3),\ \ (Danny, 8)\ \}$
2. $R = \{\ (Danny, 7),\ \ (Pete, 3),\ \ (Pete, 5)\ \}$
3. $R = \{\ (Danny, 7),\ \ (3, Pete),\ \ (Pete, 5)\ \}$

Solution

1. Even though R is a set of ordered pairs, it is not a relation from A into B , for indeed the pair $(Danny, 8)$ of R fails the test: $8 \in B$.

2. The set R is a relation from A into B . For, R is a set of ordered pairs, and every pair in R is such that its first item belongs to A , and its second item to B .

3. Even though R is a set of ordered pairs, R is not a relation from A into B . For Indeed, the pair $(3, Pete)$ of R fails to satisfy the conditions: $3 \in A$ and $Pete \in B$. ♥

Example 5: Let us suppose $A = \{\ Pete,\ Danny,\ Ann,\ \ Nicole\ \}$, $B = \{7, 3, 5\}$ and R is the relation from A into B defined by:

$$R = \{\ (Pete, 3),\ \ (Danny, 7),\ \ (Pete, 5),\ \ (Ann, 3)\ \}.$$

Then indicate which of the following propositions are true

1. $Danny\, R\, 5$ 2. $Danny\, R\, 7$ 3. $Pete\, R\, 5$
4. $Ann\, R\, 7$ 5. $Ann\, R\, 3$

Solution

1. $Danny\, R\, 5$ is false, since the ordered pair $(Danny,\ 5) \notin R$.
2. $Danny\, R\, 7$ is true, since the ordered pair $(Danny,\ 7) \in R$.
3. $Pete\, R\, 5$ is true, since the ordered pair $(Pete,\ 5) \in R$.
4. $Ann\, R\, 7$ is false, since the ordered pair $(Ann,\ 7) \notin R$.
5. $Ann\, R\, 3$ is true, since the ordered pair $(Ann,\ 3) \in R$. ♥

We now introduce the notions of the domain and range of a relation.

Definition 6 : Let R be a relation from a set X into a set Y. Then:

1. The set
$$\{ x \in X : \ \exists y \in Y, \ (x, y) \in R \}$$
is called the **domain** of R , and will be denoted by $Dom(R)$.

2. The set
$$\{ y \in Y : \ \exists x \in X, \ (x, y) \in R \}$$
is called the **range** of R , and will be denoted by $Ran(R)$.

Remark 7: In other words, the domain of a relation is simply the set of all first elements of the ordered pairs in the relation. And similarly, the range of a relation is the set of all second elements of the ordered pairs in that relation. ■

Example 8: Let us suppose $A = \{ Pete,\ Danny,\ Ann,\ Nicole \}$, $B = \{7, 3, 5\}$ and R is the relation from A into B defined by:
$$R = \{ (Pete, 3),\ (Danny, 7),\ (Pete, 5),\ (Ann, 3) \}.$$

Give $Dom(R)$ and $Ran(R)$.

Solution

$Dom(R) = \{ Pete,\ Danny,\ Ann \}$, and $Ran(R) = \{ 7,\ 3,\ 5 \}$. ♥

13.2. TYPES OF RELATIONS

In this section, we will define four important types of relations. However, before doing so, we need to introduce the simple idea of a relation **in** a set X.

Definition 9: A relation R **from** a set X **into** a set Y is said to be a relation **in** the set X, provided $X = Y$.

We are now in the position to define the following types relations: **reflective, symmetric, antisymmetric** and **transitive** relations.

REFLEXIVE RELATIONS

Definition 10: A relation R in a set X is called **reflexive**, if

$$\forall x \in X, \quad x R x$$

Example 11: State the denial of Definition 10.

Solution:
A relation R in a set X is **not** reflexive, if

$$\exists x \in X, \ \sim [x R x].$$

That is, a relation R in a set X is **not** reflexive, if

$$\exists x \in X, \ (x, x) \notin R$$. ♥

Example 12: $A = \{ 1,\ 2,\ 3,\ 4 \}$. Tell which of the following relations in A are reflexive.

1. $R = \{ (1,1),\ (2,2),\ (3,3),\ (4,4) \}$

2. $R = \{(1,1), (1,2), (2,2), (2,3), (3,3), (4,4)\}$

3. $R = \{ (1,3), (1,4), (2,2), (2,3), (2,4), (3,4), (4,4) \}$

Solution:

1. Note that for each $x \in A$, the ordered paired (x, x) belongs to R. Thus, R is a reflexive relation.

2. Here too, R is a reflexive relation. Indeed, it is easy to verify that for every item $x \in A$, the ordered paired (x, x) does belongs to R.

3. In this case, R is not a reflexive relation; for there is at least one element of A, say 1, such $(1,1) \notin R$. ♥

SYMMETRIC RELATIONS

We now give the definition of a symmetric relation.

Definition 13: A relation R in a set X is called **symmetric,** if

$$\forall x \in X,\ \forall y \in X,\quad xRy \to yRx.$$

Example 14: State the denial of Definition 13.

Solution:

A relation R in a set X is **not** symmetric, if

$$\exists x \in X,\ \exists y \in X,\quad \sim [xRy \to yRx].$$

That is, a relation R in a set X is **not** symmetric, if

$$\exists x \in X,\ \exists y \in X,\quad xRy \text{ and } \sim [yRx]\ .$$ ♥

Example 15: $A = \{ a, b, c, d \}$. Tell which of the following relations in A are symmetric.

1. $R = \{ (a,a), (b,c), (c,b), (c,c) \}$

2. $R = \{ (a,a), (a,b), (c,d), (c,c)\ (b,a) \}$.

Solution:

1. Relation R is symmetric. Indeed, by checking one by one every single ordered pair $(x, y) \in R$, we also notice that $(y, x) \in R$.

2. In this case, the relation R is not symmetric, since there is at least one ordered pair, say (c, d), which belongs to R, even though (d, c) does not in turn belong to R. ♥

ANTISYMMETRIC RELATIONS

Next, we introduce the idea of an antisymmetric relation.

Definition 16: A relation R in a set X is called **antisymmetric**, if

$$\forall x \in X,\ \forall y \in X,\ \ [xRy \text{ and } yRx] \rightarrow [x = y].$$

Remark 17: Note from Definition 16 and the solution to Example 14 that a not symmetric relation and an antisymmetric relation are two different things. ■

Example 18: State the denial of Definition 16.

Solution:

A relation R in a set X is **not** antisymmetric, if

$$\exists x \in X,\ \exists y \in X,\ \ [xRy \text{ and } yRx] \text{ and } \sim[x = y]$$

That is, a relation R in a set X is **not** antisymmetric, if

$$\exists x \in X,\ \exists y \in X,\ \ [xRy \text{ and } yRx] \text{ and } [x \neq y].$$ ♥

Example 19: Let $A = \{ a,\ b,\ c,\ d \}$, where a, b, c, and d are distinct objects. Tell which ones of the following relations in A are antisymmetric.

1. $R = \{ (a,a),\ (b,c),\ (c,b),\ (c,c) \}$

2. $R = \{ (a,a),\ (c,d),\ (c,c)\ (b,a) \}$

Solution:

1. The relation R is not antisymmetric because there are items in A, say b and c, such that bRc and cRb, even though $c \neq b$.
2. In this case, the relation R is antisymmetric. Indeed, checking all cases, we notice that for all $x \in A$, and for all $y \in A$, if xRy and yRx, then $x = y$. ♥

Example 20: Show that the relation R defined in $\mathbb{R}$ by the link phrase

... is greater than or equal to ...

is antisymmetric.

Solution:
Let $x \in \mathbb{R}$ and $y \in \mathbb{R}$, then all we need to show is that

if xRy and yRx, then $x = y$.

Suppose xRy and yRx. This means $x \geq y$ and $y \geq x$. And thus, by invoking the result of Exercise 38.3 of Chapter 9, we do obtain $x = y$. ♥

TRANSITIVE RELATIONS

Now, we will define our last relation.

Definition 21: A relation R in a set X is called **transitive**, if

$$\forall x \in X,\ \forall y \in X,\ \forall z \in X,\ [xRy \text{ and } yRz] \rightarrow [xRz].$$

Example 22: State the denial of Definition 21.

Solution:
A relation R in a set X is **not** transitive, if

$\exists x \in X,\ \exists y \in X,\ \exists z \in X,\ [xRy \text{ and } yRz] \text{ but } \sim[xRz]$. ♥

Example 23: Let $A = \{a, b, c, d\}$, where a, b, c, and d are distinct objects. Tell which ones of the following relations in A are transitive.

1. $R = \{(a,a), (b,c), (c,b), (c,c)\}$.
2. $R = \{(a,a), (c,d), (c,c)\}$.

Solution:

1. The relation R is not transitive because there are items in A, say b and c, such that bRc and cRb, even though $\sim[bRb]$.
2. In this case, the relation R is transitive because it is easy to check that for all $x \in A$, for all $y \in A$, and for all $z \in A$,

   if xRy and yRz, then xRz . ♥

Example 24: Show that the relation R defined in $\mathbb{R}$ by the **link** phrase

... is greater than ...

is transitive.

Solution:

Let $x \in \mathbb{R}$, $y \in \mathbb{R}$, and $z \in \mathbb{R}$, then all we need to show is that

if xRy and yRz, then xRz .

Suppose xRy and yRz. This means $x>y$ and $y>Z$. Therefore, by Axiom 17 of Chapter 9, we see that $x>y$, which means xRz . ♥

13.3. ORDER RELATIONS

Some relations, such as "$\geq$", can be used to order the elements of sets in which they are defined. These relations are simultaneously reflexive, antisymmetric and transitive. Due to their great importance in mathematics and in everyday life, these relations have been given the special name of **ordered relations**.

> **Definition 25 [Order relations]:** A relation R in a set X is called an **order relation,** if R is reflexive, antisymmetric and transitive.

Example 26: Show that the relation R defined in $\mathbb{N}$ by

$$\text{“} rRs, \quad \textit{if } r \textit{ is divisible by } s \text{”}$$

is an order relation. [The reader is reminded that a natural number r is divisible by a natural number s, if there is a third natural number k such that $r = ks$.]

Solution:

Step 1. The relation R *is reflexive*: Indeed, for every natural number r, we know that $r = 1 \cdot r$. Thus rRr.

Step 2. The relation R *is antisymmetric:* Indeed, suppose r and s are any two natural numbers satisfying the conditions rRs and sRr; we will show that this supposition implies that $r = s$.

On the one hand,

$$rRs \quad \Rightarrow \quad \exists k \in \mathbb{N}, \quad r = ks. \tag{1}$$

On the other hand,

$$sRr \quad \Rightarrow \quad \exists l \in \mathbb{N}, \quad s = lr. \tag{2}$$

Therefore, plugging (1) into (2), we have

$$1 = kl.$$

But, k and l being positive whole numbers, we obtain

$$k = 1 = l \tag{3}$$

Hence, by plugging (3) into (1) or (2) we have

$$s = r\,.$$

Step3. The relation R is transitive: Indeed, suppose r , s and t are any three natural numbers satisfying the conditions $r\,R\,s$ and $s\,R\,t$; we will show that this supposition implies that $r\,R\,t$. On the one hand,

$$rRs \;\;\Rightarrow\;\; \exists k \in \mathbb{N}, \;\; r = k\,s\,. \qquad (4)$$

On the other hand,

$$sRt \;\;\Rightarrow\;\; \exists l \in \mathbb{N}, \;\; s = l\,t\,. \qquad (5)$$

Therefore, by plugging (5) into (4), we have

$$r = (kl)t\,, \quad \textit{with} \;\; kl \in \mathbb{N}\,.$$

Thus r is divisible by t . That is to say $r\,R\,t$. ♥

Definition 27 [Comparable elements]: Let X be a set equipped with an order relation R . And assume a and b are elements of X . Then, a and b are said to be **comparable** by R , if at least one of these two sentences is true.

1. $a\,R\,b$

or

2. $b\,R\,a$

If every pair of elements in X is comparable by R , we call R a **total order** in X ; otherwise R is called a **partial order** in X .

Example 28: Show that the order relation of Example 26 is a partial order.

Solution:

All that one has to do is to come up with one pair of natural numbers that is not comparable by the relation "…*is divisible by... .*" Since this is a trivial task, we leave it to the reader to supply all the details . ♥

Example 29: Show that the "*...less than or equal to...*" relation is a total order in the set $\mathbb{R}$ of real numbers.

Solution:
Since this too is a trivial task, we leave this too to the reader to supply all the details . ♥

13.4. EQUIVALENT RELATIONS

In this section we will define another special kind of relations called equivalent relations. Equivalent relations are among the most important concepts in modern mathematics.

Definition 30 [Equivalent relations]: A relation R in a set X is called an **equivalent relation**, if R is reflexive, symmetric and transitive.

Remark 31: Note the difference between order relations and equivalent relations. The difference is that order relations must be antisymmetric; whereas equivalent relations have to be symmetric . ■

Example 32: Verify that the "*...equal to...*" relation is an equivalent relation.

Solution:
Probably, one of the most utilized relations in mathematics is the *equal to* relation denoted by " = ." Indeed, much of mathematics involves the study of equations, the basis of which resides in the fact that the "*...equal to...*" relation is [by Axioms 1, 2, and 3 of Chapter 9] an equivalent relation:

- Reflexivity:

$$\forall x \in X, \quad x = x.$$

- Symmetry:

$$\forall x \in X, \quad \forall y \in X, \quad [x = y] \rightarrow [y = x].$$

- Transitivity:

$$\forall x \in X, \forall y \in X, \forall z \in X, [x = y \text{ and } y = z] \rightarrow [x = z] .$$ ♥

13.5. SET PARTITIONING

DEFINITION OF A SET PARTITION

In the preceding example, we mentioned the relation of equality which most people would consider a trivial case of equivalence relations. Hence, in our next theorem we will provide a less obvious one. To be able to do so, we need to introduce one more definition.

Definition 33 [Partition of a set]: Let X be a nonempty set, and $\mathbb{P}$ a collection of subsets of X. Then, $\mathbb{P}$ is said to be a **partition** of X, if

1. Each element of $\mathbb{P}$ is a nonempty subset of X,

and

2. Every object in X belongs to one and only one element of $\mathbb{P}$.

Remark 34: It is absolutely important that we understand that a partition is a set whose members are themselves sets. However, we will prefer the use of the phrase *collection of subsets* to the expression *set of subsets*, which is monotonous if not cacophonic.

Example 35: Give three different partitions of the set $X = \{ a, b, c, d \}$.

Solution:

According to Definition 33, each of the following collections of sets qualifies as a partition of X:

$$\mathbb{P} = \{X\} = \{ \{ a, b, c, d \} \}$$

$$\mathbb{P} = \{ \{ a\}, \{b, c, d \} \}$$

$$\mathbb{P} = \{ \{ a, b\}, \{ c, d \} \} .$$ ♥

Example 36: Let X be a nonempty set. Explain why 2^X is not a partition of X .

Solution:

The power set 2^X is not a partition of X , since it contains the empty set $\varnothing$. Indeed, by Definition 33, every element of a partition is supposed to be a nonempty set. ♥

Theorem 37: Let $\mathbb{P}$ be a given partition of a set X , and R the relation defined in X by:

$$xRy,\ \textit{if both } x \textit{ and } y \textit{ belong to the same set } S \in \mathbb{P}.$$

Then, R is an equivalent relation.

Proof:

Part 1. Reflexivity: Let x be in X . Then, we know from the definition of a partition that there is a unique set $S \in \mathbb{P}$ such that $x \in S$. Thus, x and x belong to the same set $S \in \mathbb{P}$, Hence $x R x$.

Part 2. Symmetry: Let x and y be elements of X such $x R y$. This means that both x and y belong to the same unique set $S \in \mathbb{P}$. Thus, both y and x belong to the same set S . Hence, $y R x$.

Part 3. Transitivity: Let x , y and z be elements of X such $x R y$ and $y R z$. On the one hand,

$$x R y \;\Rightarrow\; x \textit{ and } y \in A, \textit{ for some set } A \in \mathbb{P}. \tag{6}$$

On the other hand,

$$y R z \;\Rightarrow\; y \textit{ and } z \in B, \textit{ for some set } B \in \mathbb{P}. \tag{7}$$

Also, we do know from the definition of a partition that y must belongs to one and only one element of $\mathbb{P}$. Hence, $A = B$. Therefore, both x and z belong to the same set $A \in \mathbb{P}$. Thus, $x R z$. ♣

Example 38: Consider the set

$$X = \{ a, b, c \}$$

with the partition

$$\mathbb{P} = \{ \{a\}, \{b, c\} \}$$

List the equivalent relation R defined by this partition $\mathbb{P}$.

Solution

$$R = \{ (a,a), (b,b), (c,c), (b,c), (c,b) \} . \heartsuit$$

EQUIVALENT CLASSES

With the preceding theorem, we were able to show that every partition of a set X determines in X an equivalent relation. We now want to prove something of a converse of that theorem. More precisely, we want to show that by equipping a set X with an equivalent relation, we are also implicitly defining on it a partition.

To do this, we will first re-state Definition 33 in a form that we can readily used here.

Definition 39 [Partition of a set]: Let X be a nonempty set, and $\mathbb{P}$ a collection of subsets of X . Then, $\mathbb{P}$ is called a **partition** of X , if

1. If $A \in \mathbb{P}$, then $A \neq \varnothing$,
2. If $A \in \mathbb{P}$ and $B \in \mathbb{P}$, then either $A = B$ or $A \cap B = \varnothing$,

and

3. If $x \in X$, then there is $A \in \mathbb{P}$ such that $x \in A$.

Remark 40: It is a trivial task to check that Definitions 33 and 39 are equivalent. That is, if you have either definition, then you have the other one too. We therefore leave it to the reader to logically convince him- or herself that this is indeed the case. ■

We are now in the position to present and prove the result mentioned at the opening of this subsection.

Definition and Theorem 41: Let R be an equivalent relation in a nonempty set X. Furthermore, for each $a \in X$, let us denote by $cl(a)$ the set,

$$cl(a) = \{ x \in X : xRa \},$$

of all elements of X that are related to a. We call $cl(a)$ the **equivalent class** of the element a.

Then the collection,

$$C = \{ cl(a) : a \in X \},$$

of all equivalent classes of elements of X constitutes a partition of X.

Proof:
In light of Definition 39, all we need to show is three things:

First thing. We should prove that for every $a \in X$, $cl(a) \neq \varnothing$:
For each $a \in X$, we know that aRa since R is an equivalent relation. Therefore, $a \in cl(a)$. Thus, $cl(a)$ is indeed nonempty.

Second thing. We need to show that for every $a \in X$ *and* $b \in X$, *either* $cl(a) \cap cl(b) = \varnothing$ or $cl(a) = cl(b)$:
Suppose $a \in X$, *and* $b \in X$. Then, we have only two cases:

Case1. $cl(a) \cap cl(b) = \varnothing$. In this case, because $cl(a)$ and $cl(b)$ are each nonempty, the equality $cl(a) = cl(b)$ cannot hold.

Case 2. $cl(a) \cap cl(b) \neq \varnothing$. Then, there is at least one element, say $c \in X$, which belongs to both $cl(a)$ and $cl(b)$. Hence:

$$aRc \quad \text{and} \quad cRb,$$

which, given the fact that R is transitive, in turn means that

$$aRb \quad . \tag{8}$$

Therefore, using the transitivity of R again, we see that every element of X that is related to a is also related to b, and vice versa. Thus, $cl(a) = cl(b)$.

Third thing. We must show that if $x \in X$, *then* $x \in cl(a)$ *for some* $a \in X$: This is pretty obvious. Indeed, for every $x \in X$, all we need to do is to pick $a = x$, and we automatically will have $x \in cl(a)$. ♣

Example 42: Consider the set

$$X = \{ a, \ b, \ c, \ d \}$$

equipped with the relation

$$R = \{ (a,a), (b,b), (c,c), (a,d), (d,c), (a,c), (c,a), (d,a), (c,d), (d,d)\}.$$

1. Check that R is an equivalent relation in X.
2. Determine the partition C associated with the equivalent relation R.

Solution

1. Verification of : R is an equivalent relation.

Reflexivity: Since (a,a), (b,b), (c,c) and (d,d) all belong to R, the relation R is reflexive.

Symmetry: By checking all ordered pairs, we notice that for every $(x,y) \in R$, we also do have $(y,x) \in R$. Thus the relation R is symmetric.

Transitivity: By examining all ordered pairs, we see that for every elements x, y and z in X, if $(x,y) \in R$ and $(y,z) \in R$, then so is $(x,z) \in R$. Thus R is transitive.

2. It is not difficult to see that

$$cl(a) = \{a, d, c\}, \qquad cl(b) = \{b\},$$
$$cl(c) = \{a, d, c\}, \qquad \text{and} \quad cl(d) = \{a, d, c\}.$$

Thus, the partition C associated with the equivalent relation R is

$$C = \{ \, cl(a) : \, a \in X \, \} = \{ \, \{b\}, \, \{a, \, c, \, b\} \} .$$

Check that C is indeed a partition of the set X . ♥

13.6. CHAINS AND CYCLES

In this section we will discuss two new concepts; the idea of a **chain** and that of a **cyclic relation**.

Definition 43 [Chain]: Let R be a relation in a set X . By **a chain** of length $n \in \mathbb{N}$ we mean any ordered set $(a_1, \, a_2, \, a_3, \, . \, . \, . \, , \, a_n)$[15] of elements of X such that

$$\forall i \in \{1, \, 2, \, 3, . \, . \, ., n-1\}, \quad a_i \, R \, a_{i+1} \, .$$

Example 44: Consider the set

$$X = \{ \, a, \; b, \; c, \; d \, \}$$

equipped with the relation

$$R = \{ \, (a,a), \, (b,b), \, (c,c), \, (d,d), \, (a,d), \\ (d,c), \, (a,c), \, (c,b), \, (d,a), \, (c,d)\} \, .$$

1. Give a chain of length 2.
2. Give a chain of length 4.

Solution

1. $(a, \, d)$ is a chain of length 2, since it is made up of two items, and $a \, R \, d$.
2. $(a, \, d, \, c, \, b)$ is a chain of length 4, since on the one hand it is made up of four items, and on the other hand we have $a \, R \, d$, $d \, R \, c$ and $c \, R \, b$. ♥

[15] 15 We want the reader to note that , like an ordered pair, an ordered set allows repetition in the listing of its elements

Definition 45 [Cycle]: Let R be a relation in a set X , and assume the ordered list $(a_1, a_2, a_3, \ldots, a_n)$ is a chain of elements of X. Then $(a_1, a_2, a_3, \ldots, a_n)$ is said to be a **cyclic chain** or simply **cycle**, if

1. the a_i are all distinct.

and

2. $a_n R a_1$

Example 46: Consider the set

$$X = \{ a, b, c, d \}$$

equipped with the relation

$$R = \{ (a,a), (b,b), (c,c), (d,d), (a,d), (d,c), (a,c), (c,b), (b,a), (c,d)\}.$$

Find in R a cycle of length equal to 4.

Solution

The ordered set (a, d, c, b) is a cyclic chain, since on the one hand all of its elements are distinct, and on the other han aRd, dRc and cRb and bRa . ♥

Theorem 47: Let R be a reflexive and transitive relation in a nonempty set X. Then:

R is antisymmetric $\Leftrightarrow$ Every cycle of X is of length 1.

Proof:

We need to show two things:

1. R is antisymmetric $\Rightarrow$ Every cycle of X is of length one,

and

2. Every cycle of X is of length one $\Rightarrow$ R is antisymmetric.

The proofs of these two implications are next:

1. We will prove the first part by **contradiction**. Assume that R is antisymmetric, but that X contains a cycle $(a_1, a_2, a_3, \ldots, a_n)$ of length $n \geq 2$. Thus,

$$a_n R a_1 \,. \tag{9}$$

Also, by the transitivity of R and the fact that $a_i R a_{i+1}$ for all $i \in \{1, 2, 3, \ldots, n-1\}$, we have

$$a_1 R a_n \,. \tag{10}$$

Hence, combining (9) and (10) with the antisymmetry of R, we deduce that $a_n = a_1$; which contradicts the fact that all the elements of the cycle $(a_1, a_2, a_3, \ldots, a_n)$ must be distinct.

2. To show the second part of the theorem, let us suppose that every cycle of X is of length one. Now, we should prove from this supposition that R is antisymmetric.

Let a and b be any two elements of X such that $a R b$ and $b R a$. This means that (a, b) is a cycle. Therefore, in view of the supposition that every cycle must be of length one, we conclude that $a = b$, which completes the proof. ♣

Remark 48: The meaning of this theorem is twofold: On the one hand, it says that in a set with an order relation there is no chain whose length exceeds 1. On the other hand, it shows that if a set equipped with a reflexive and transive relation contains no cycle of length greater than 1, then its relation is an order relation. ■

PROBLEMS

1. Earlier in the beginning of this chapter we introduced the concept of an ordered pair somehow informaly; now we will give a more formal [if not a more rigorous] definition based on the idea of a set. Let a and b be any two elements of a nonempty set X ; we now define the ordered pair (a, b) as follows:

$$(a, b) = \{ a, \{ a, b \} \}.$$

Using this definition, show that

$$(a, b) = (x, y) \Leftrightarrow a = x \quad and \quad b = y.$$

2. Let $X = \{ a, x \}$.

2.1. List all the relations in X.

2.2. List those relations that are reflexive

2.3. List those relations that are transitive

2.4. List those relations that are symmetric

2.2. List those relations that are antisymmetric

2.2. List those relations that are equivalent relations

3. Let X be a nonempty set and $R \subset X \times X$ be the empty set. Which ones of these properties does R have?

3.1. Reflexivity.

3.2. Transitivity.

3.3. Symmetry.

3.4. Antisymmetry.

4. Let R and S be any two partial orders in a set X. Show that $R \cap S$ too is a partial order.

5. Prove that each of these relations is an equivalent relation.

5.1. For every $x, y \in \mathbb{R}$, $\quad x R y \Leftrightarrow x - y \in \mathbb{Z}$.

5.2. Suppose $f : A \to B$ is a given function.

$$x, y \in A, \quad x R y \Leftrightarrow f(x) = f(y).$$

5.3. For every $(a, b), (x, y) \in \mathbb{Z} \times \mathbb{Z}$,

$$(a, b) R (x, y) \Leftrightarrow a + y = b + x.$$

6. Give two different partitions of the set:

$$X = \{ Mark, \ Paul, \ Anna, \ Jean, \ Pete \}.$$

7. For each partition in Problem 6, list the equivalent relation R induced by that partition.

"The defects in the mathematical training of the student of engineering appear to be largely in the knowledge and grasp of the fundamental principles; and the constant effort of the teacher should be to ground the student thoroughly in the fundamentals, which are too often lost sight of in the mass of details."
American Mathematical Monthly. 18, (1911) p.24

CHAPTER 14

FUNCTIONS AND ABSOLUTE VALUE

14.1. FUNCTIONS

In the last chapter we studied relations. In the present one we will discuss functions; a function is a special type of relation. The importance of functions can hardly be overemphasized; indeed, the concept pervades the entire field of modern mathematics. From algebra through geometry to calculus, one cannot avoid the use of functions.

Definition1 [Function]: A function f defined **on** a set X **into** a set Y is simply any relation from X into Y that satisfies these two conditions:

1. $\text{Dom}(f) = X$,

and

2. $(x, y_1) \in f$ and $(x, y_2) \in f \Rightarrow y_1 = y_2$.

Remark 2: To put it in plain English, the first condition means that every element of set X must be related to some element of set Y. As for the second condition, it simply requires that every element x of X be related to one and only one element of Y. ■

Thus, a function may also be defined as follows

Definition 3 [Function]: A function f **on** a set X **into** a set Y is a rule[16] that pairs each element x of X with exactly one element y in Y. Set X is called the **domain** of the function f, and set Y is its **codomain**.

Remarks 4:

1. For any element $x \in X$, it is customary to denote by $f(x)$ the unique item in Y that is paired with x. It is also customary to read $f(x)$ in one of three ways:

   (i). f of x,
   (ii). the image of x by f,
   (iii). or the value of f at x.

Therefore, one should be careful not to confuse $f(x)$ with f; whereas f stands for the function itself, the notation $f(x)$ as explained above is the image of a typical $x \in X$.

2. The variable $x \in X$ in the notation $f(x)$ is called the **argument** or the **input** [or the **independent variable**, since its value can be chosen arbitrarily from X]. In contrast, we sometimes let $y = f(x)$ and call $y \in Y$ the **dependent variable** of the function f; this is because the value of y is then clearly dependent on the choice of x.

3. Sometimes, we will find it a convenient shorthand to write $f: X \to Y$ to indicate that f is a function defined on the set X with values into set Y. ■

[16] Apart from the requirement that the rule f must pair every element of X with one and only one element of Y, there is no further restriction whatsoever placed on f.

Example 5: We know from **the First Archimedean Principle** [Theorem 30 of Chapter 11] that for every real number x , there is one and only one integer k_0 such that:

$$k_0 \leq x < k_0 + 1 \qquad (1)$$

Thus, the rule

$$E: \mathbb{R} \to \mathbb{R} \quad \text{defined by} \quad E(x) = k_0, \ \forall x \in \mathbb{R},$$

with k_0 given by (1), is a function. Note that the function E pairs every $x \in \mathbb{R}$ with the biggest integer less than or equal to x. For this reason, the function E is called the **floor function**, and $E(x)$ the **integer part** of x . ♥

Example 6: Let S be any nonempty set. We shall denote by I the rule which assigns to each x in S the object x itself. Then, clearly,

$$I: S \to S$$

is a function. The function I is called the **identity function** on set S . ♥

Note that for a function f to be completely and uniquely determined, three things must be specified in advance: its domain D_f , its codomain C_f , and the rule that gives the image $f(x)$ *for any* $x \in D_f$. Hence:

Definition 7 : Two functions $f: D_f \to C_f$ and $g: D_g \to C_g$ are said to be **equal**, if

1. their domains are equal

$$D_f = D_g,$$

2. their codomains are equal

$$C_f = C_g,$$

and

3. they obey the same rule

$$f(x) = g(x), \qquad \forall x \in D_f .$$

14.2. THE ABSOLUTE VALUE FUNCTION

We shall now discuss the function called the **absolute value function**. The importance of this function in mathematics can hardly be overemphasized. Indeed, this function and some of its properties will be invoked several times in the remainder of this book.

Despite its reputation as a confusing concept to many high-schoolers, the absolute value function should not be feared. It should be perfectly intelligible to anyone who keeps in mind the fact that a function is essentially a rule.

Think of a rule $f:\mathbb{R}\to\mathbb{R}$ that behaves as follows. If x is a negative number, then the rule f assigns to x its opposite $-x$; but if x is nonnegative [that is to say, positive or equal to zero], then f assigns to x the number x itself. Thus, clearly, for every $x\in\mathbb{R}$, we may write $f(x)=-x,\ \textit{if}\ x<0$ and $f(x)=x,\ \textit{if}\ x\geq 0$, or more compactly:

$$f(x)=\begin{cases}-x, & \textit{if}\ x<0,\\ x, & \textit{if}\ x\geq 0.\end{cases}$$

The function f , we have just defined, is precisely what is referred to as the **absolute value function**. And the image $f(x)$ of a typical real number x, also commonly denoted by $|x|$, is called the **absolute value** of x. Hence,

$$|x|=\begin{cases}-x, & \textit{if}\ x<0,\\ x, & \textit{if}\ x\geq 0.\end{cases}$$

Example 7: Rewrite the following expressions without the absolute value symbol "$|\ |$".

1. $|5|$, 2. $|-5|$, 3. $|-3x+6|$, 4. $|x^2-x|$.

Solution

1. $|5| = 5$, because $5 \geq 0$.

2. $|-5| = -(-5) = 5$, because $-5 < 0$.

3. *We have exactly two cases here*: The expression $-3x+6$, within the absolute value symbol, can be either negative or nonnegative depending on the value of its variable x.

   Case 1. If $-3x+6<0$; that is to say if $x \in]2,\ +\infty[$, then

   $$|-3x+2| = -(-3x+6) = 3x-6$$

   Case 2. If $-3x+6 \geq 0$; that is to say if $x \in]-\infty,\ 2]$, then

   $$|-3x+6| = -3x+6.$$

4. Here too we will have two cases, since $x^2 - x$ can be either negative or nonnegative.

   Case 1. If $x^2 - x \geq 0$; that is to say if $x \in]-\infty,\ 0] \cup [1,\ +\infty[$ [17], then

   $$|x^2 - x| = x^2 - x$$

   Case 2. If $x^2 - x < 0$; that is to say if $x \in]0,\ 1[$, then

   $$|x^2 - x| = -(x^2 - x) = -x^2 + x\ . ♥$$

Next, we present some simple but important properties of the absolute value function.

[17] Refer to Example 14 of Chapter 5, in case you need to know how we arrived at $x \in]0,\ 1[$

Theorem 8: For every x and y in $\mathbb{R}$,

1. $|x| \geq 0$,
2. $-|x| \leq x \leq |x|$,
3. $|xy| = |x|\,|y|$,
4. $\left|\frac{x}{y}\right| = \frac{|x|}{|y|}$.

Proof: Each of the four parts of this theorem is easily established by the **method of exhaution**. Hence, we will prove part 1 only, and leave the proofs of the three other parts to the reader.

1. There are exactly two cases here.

   Case 1. $x \geq 0$: Then $|x| = x \geq 0$.

   Case 2. $x < 0$: Then $|x| = -x \geq 0$, by virtue of Theorem 36.1 of Chapter 9. ♣

We are now going to prove probably the most important property of the absolute value function. We will use this property and its corollaries [or consequences] in Chapters 16 and 17.

Theorem 9: For every $x \in \mathbb{R}$, and every real number $M > 0$,

$$|x| \leq M \Leftrightarrow -M \leq x \leq M$$

Proof: The proof of this theorem consists of two parts.

Part 1. Proof of $|x| \leq M \Rightarrow -M \leq x \leq M$. We will do this by the **method of exhaustion.** In other words, we will examine all possible cases:

Case 1. $x \geq 0$: In this case, we know that $|x| = x$. Hence, from the obvious fact that $-M \leq |x| \leq M$, we automatically deduce that $-M \leq x \leq M$.

Case 2. $x < 0$: In this case, we know that $|x| = -x$. Hence, from the fact that $-M \leq |x| \leq M$, we automatically deduce that $-M \leq -x \leq M$. Thus, multiplying all three sides of this double inequality by -1 , we arrive at the desired result, that isto say $-M \leq x \leq M$.

Part 2. Proof of $-M \leq x \leq M \Rightarrow |x| \leq M$. Here too, we will do the proof by looking at all cases.

Case 1. $x \geq 0$. In this case, we know that $|x| = x$. Thus, the double inequality $-M \leq x \leq M$ may also be written as $-M \leq |x| \leq M$, which automatically gives us the single inequality $|x| \leq M$.

Case 2. $x < 0$. In this case, we know that $|x| = -x$. Hence, the double inequality $-M \leq x \leq M$ may be written as $-M \leq -|x| \leq M$ which, after multiplying all three sides by -1 , yields the desired result $|x| \leq M$. ♣

Next, we give three important consequences of the preceding theorem.

Corollary 10: For every x and y in $\mathbb{R}$,

1. $|x+y| \leq |x| + |y|$, **The triangle inequality,**
2. $|x-y| \leq |x| + |y|$,
3. $||x| - |y|| \leq |x-y|$.

Proof:

1. By virtue of Theorem 9, to prove that $|x+y| \le |x|+|y|$ holds, we will just need to establish the double inequality $-(|x|+|y|) \le x+y \le |x|+|y|$.

We know from Theorem 8.2 that

$$-|x| \le x \le |x|, \tag{2}$$

for every $x \in \mathbb{R}$.

Similarly, for every $y \in \mathbb{R}$, we have

$$-|y| \le y \le |y|. \tag{3}$$

Thus, by virtue of the result of Problem 6 of Chapter 10, adding the two inequalities (2) and (3) sidewise, we obtain

$$-(|x|+|y|) \le x+y \le |x|+|y|,$$

which completes the proof of the first part of this theorem.

2. As for this part of the theorem, we will show using the method of **Direct Proofs** that it is a straightforward consequence of Part 1. Indeed, for all x and y in $\mathbb{R}$,

$$\begin{aligned} |x-y| &= |x+(-y)|, && \text{by the definition of subtraction} \\ &\le |x|+|(-y)|, && \text{by Part 1 of this theorem} \\ &= |x|+|y|, && \text{because } |-y|=|y| \text{; prove it.} \end{aligned}$$

Thus, we have shown that $|x-y| \le |x|+|y|$ for all x and y in $\mathbb{R}$.

3. This part of the theorem too may be shown by using the method of **Direct Proofs**. Indeed, for all x and y in $\mathbb{R}$, we have,

$$\begin{aligned} |x| &= |(x-y)+y| \\ &\le |x-y|+|y|, && \text{by Part 1 of this theorem.} \end{aligned}$$

Hence, $|x| \leq |x-y|+|y|$, for all x and y in $\mathbb{R}$. Therefore, for all x and y in $\mathbb{R}$, we have

$$|x|-|y| \leq |x-y| \tag{4}$$

Now, notice that we can swap x and y in (4), since the two numbers are completely arbitrary. Doing so, gives us

$$|y|-|x| \leq |y-x|,$$

which, by multiplying by -1, yields

$$-|x-y| \leq |x|-|y|. \tag{5}$$

Therefore, combining (4) and (5), we get the double inequality

$$-|x-y| \leq |x|-|y| \leq |x-y| .$$

Hence, by Theorem 9, we deduce that

$$||x|-|y|| \leq |x-y| .$$

And the entire proof is complete. ♣

14.3. A USEFUL INTERPRETATION OF THE ABSOLUTE VALUE FUNCTION

There is one important interpretation of absolute values that is worth mentioning, if we are to form a useful intuitive understanding of the concept. We introduce this interpretation as

Definition 11 [Distance]: Let x and y be any two real numbers. Then,

$$|x-y| ,$$

the absolute value of the difference between x and y, is the **distance** between x and y. We shall denote it by $d(x,y)$.

Remark 12: Notice that by making $y = 0$ in the foregoing definition, we obtain

$$d(x,0) = |\, x \,| \, . \tag{6}$$

Thus, the absolute value of any real number x is simply the distance [on the number line] from the origin 0 to the number x . ■

Example 13: Rewrite the distance formula

$$d(x,y) = |\, x - y \,|$$

without explicitly using the absolute value symbol.

Solution

From the definition of absolute value, we have

$$|\, x - y \,| = \begin{cases} x - y, & \textit{if } x - y \geq 0, \\ -(x - y), & \textit{if } x - y < 0. \end{cases}$$

Thus,

$$d(x,\, y) = \begin{cases} x - y, & \textit{if } x \geq y, \\ y - x, & \textit{if } y > x. \end{cases}$$

Remark 14: In other words, the distance between any two numbers x and y may be obtained by simply subtracting the smaller number from the bigger one. **Hence, a distance is never a negative numbers .** ■

Example 15: Assume ε is a fixed number in $\mathbb{R}$. Then, depending on the sign of ε , determine the set:

$$A = \left\{ x \in \mathbb{R} : \ |\, x \,| = \varepsilon \right\} .$$

Solution

There are three cases depending on the sign of ε :

Case 1. $\varepsilon < 0$ **:** In this case, clearly $A = \varnothing$, the empty set. For, indeed, by Theorem 8.1, there is no such a real number x whose absolute value [or distance from the origin 0] is negative.

Case 2. $\varepsilon = 0$ **:** In this case, $A = \{x \in \mathbb{R} : |x| = 0\}$. Thus, A is the **singleton**[18] $A = \{0\}$; for 0 is the only real number whose distance from the origin is 0.

Case 3. $\varepsilon > 0$ **:** In this case, we have

$$A = \{x \in \mathbb{R} : |x| = \varepsilon\} = \{x \in \mathbb{R} : x = \varepsilon, \text{ or } x = -\varepsilon\},$$

since there are exactly two real numbers, namely ε and $-\varepsilon$, whose distances from the origin equal ε **.** ♥

14.4. COMPOSITION OF FUNCTIONS

In this section, we will introduce the important concept of a **composite function**.

Assume we are given two functions

$$f : D_f \to C_f$$

and

$$g : D_g \to C_g \quad .$$

Then, for every $x \in D_f$ we can determine $f(x)$ using the first function. Moreover, if it turns out that $f(x) \in D_g$, then we may also determine $g[f(x)]$ using the second function. In other words, we may be able to **compose** or **combine** the two functions into a new one that assigns every element x of a subset of D_f with the unique item $g[f(x)]$ in set C_g.

---

[18] A **singleton** is a set with a single element in it.

Definition 16 [Composite function]: Given two functions g and f, the **composite function** of g and f, denoted by gof, is the new function defined by:

$$gof : \{ x \in D_f : \ f(x) \in D_g \} \to C_g$$

and

$$(gof)(x) = g[f(x)] \, .$$

The notation gof is read " g **composite** f " or simply " g **circle** f "

Remark 17: Note that, by Defition 16, $D_{gof} \subset D_f$. Also, note that, by naming U the universal set for the variable x, we have:

$$\{ x \in D_f : \ f(x) \in D_g \} = \{ x \in U : \ x \in D_f \ \ and \ \ f(x) \in D_g \} . \ \blacksquare$$

Example 18: Let $u(x) = x^2 + 1$ and $v(x) = \sqrt{x}$.

1. Find D_{uov}, 2. Find uov, 3. Find D_{vou},
4. Find vou

Solution

1. If you can remember, we learned earlier on that no function f is fully determined until and unless one knows its domain, codomain, and the rule giving the image of each $x \in D_f$. Therefore, before anything else, we need to find the domains and codomains of our two functions u and v. Note that they are not explicitly given here. In a case such as this, where only the formula $f(x)$ is explicitly given, *the domain of f is, by convention, the largest set of real numbers for which $f(x)$ in turn is a real numbers. That is:*

$$D_f = \{ x \in \mathbb{R} : \ f(x) \in \mathbb{R} \} .$$

As for the codomain C_f of such an incompletely specified function, it is customarily taken to be the entire set $\mathbb{R}$. These conventions are commonly known as the **rules of maximal domain and codomain**. Applying these rules to u and v we automatically get:

$$D_u = \mathbb{R} \quad \text{and} \quad D_v = \{x \in \mathbb{R} : \ x \geq 0\}. \qquad (7)$$

Now, using Definition 16 and Remark 17 together with (7), we find:

$$\begin{aligned} D_{uov} &= \{x \in D_v : \ v(x) \in D_u\} \\ &= \{x \in \mathbb{R} : \ x \in D_v, \ \textit{and} \ v(x) \in D_u\} \\ &= \{x \in \mathbb{R} : \ x \geq 0, \ \textit{and} \ \sqrt{x} \in \mathbb{R}\} \\ &= \{x \in \mathbb{R} : \ x \geq 0\} \cap \{x \in \mathbb{R} : \ \sqrt{x} \in \mathbb{R}\} \\ &= \{x \in \mathbb{R} : \ x \geq 0\} \cap \{x \in \mathbb{R} : \ x \geq 0\} \\ &= \{x \in \mathbb{R} : \ x \geq 0\} \\ &= [0, +\infty[. \end{aligned}$$

2. $(uov)(x) = u[v(x)] = u\left[\sqrt{x}\right] = \left[\sqrt{x}\right]^2 + 1 = x + 1.$

3. $$\begin{aligned} D_{vou} &= \{x \in D_u : u(x) \in D_v\} \\ &= \{x \in \mathbb{R} : \ x \in D_u, \ \textit{and} \ u(x) \in D_v\} \\ &= \{x \in \mathbb{R} : \ x \in \mathbb{R}, \ \textit{and} \ x^2 + 1 \geq 0\} \\ &= \{x \in \mathbb{R} : \ x \in \mathbb{R}\} \cap \{x \in \mathbb{R} : \ x^2 + 1 \geq 0\} \\ &= \mathbb{R} \cap \mathbb{R} \\ &= \mathbb{R}. \end{aligned}$$

4. $(vou)(x) = v[u(x)] = v\left[x^2 + 1\right] = \sqrt{x^2 + 1}$. ♥

Remark 19: Note that the expressions for $(u \circ v)(x)$ and $(v \circ u)(x)$ found in the foregoing example are two different algebraic expressions. In fact, in general, given distinct functions f and g, their two compositions,

$$f \circ g \text{ and } g \circ f \ ,$$

are distinct functions. Hence, composition is not commutative . ■

PROBLEMS

1. * Let A be a set. Consider the function:

$$\chi_A : A \to \{0, 1\} \quad \text{such that} \quad \chi_A(x) = \begin{cases} 1, & \text{if } x \in A \\ 0, & \text{if } x \in U - A, \end{cases}$$

where U is the universe of discourse. The function χ_A is called the **characteristic function** of the set A. Show that both χ_U and $\chi_\varnothing$ are constant functions.

2. Rewrite these expressions without the absolute value sign:

   2.1. $|2 - x|$.

   2.2. $|2x - x^2|$.

   2.3. $|x^2 - 5x - 6|$.

   2.4. $|x^3 + 3x^2 - 3x - 1|$.

3. In each case below find the maximal domain of the function f.

   3.1. $f(x) = \dfrac{1}{x+1}$

   3.2. $f(x) = \dfrac{1}{x^2 - 1}$

3.3. $f(x) = \dfrac{3x-5}{x^2-x-2}$

3.4. $f(x) = \dfrac{x+1}{x^2-x-2}$

3.5 $f(x) = u(v(x))$, *where* $v(x) = \sqrt{2x-5}$ *and* $u(x) = x^2 - \frac{3}{2}$

3.6 $f(x) = v(u(x))$, *where* $v(x) = \sqrt{2x-5}$ *and* $u(x) = x^2 - \frac{3}{2}$

4. * Show that $s_n \in]L-\varepsilon,\ L+\varepsilon[\ \Leftrightarrow\ |s_n - L| < \varepsilon$.

5. Assume we are given three functions:

$$u : X \to Y, \quad v : Y \to Z, \quad \text{and} \quad w : Z \to W$$

Show that

$$(u o v) o w = u o (v o w).$$

[In light of this associativity, we may omit the parentheses in the above, and write simply $u o v o w$ for $(v o u) o w$ or $v o (u o w)$.]

"If I were again beginning my studies, I would follow
the advice of Plato and start with mathematics."
Galilei, Galileo.

CHAPTER 15

TYPES OF FUNCTIONS

In this Chapter we will define three important types of functions encountered in all branches of modern mathematics. These three classes of functions are the **injective** functions, the **surjective** functions, and the **bijective** functions. We will also study some of their properties.

15.1. INJECTIVE FUNCTIONS

If you can remember, we learned in the preceding chapter that any function f is required to assign to every item in its domain D_f one and only one image. This means that the condition

$$\forall x \in D_f, \quad \forall y \in D_f, \quad x = y \rightarrow f(x) = f(y) \qquad (1)$$

must hold for every function f .

So, now we may ask ourselves whether the converse of (1), that is

$$\forall x \in D_f, \quad \forall y \in D_f, \quad f(x) = f(y) \rightarrow x = y \qquad (2)$$

also hold for every function f ?

Clearly, the answer to this question is a "no", since for at least the function $s(x) = x^2$, we have $s(-1) = s(1)$ even though $-1 \neq 1$.

Hence, we see that while (1) is satisfied by all functions, there are functions, such as $s(x) = x^2$, for which (2) does not hold true.

Those function for which condition (2) holds true are called **injections** or **injective functions**.

Definition1 [Injection]: A function $f : D_f \to C_f$ is an **injection**, provided

$$\forall x \in D_f, \ \forall y \in D_f, \ f(x) = f(y) \to x = y . \qquad (3)$$

Example 2: Show that the function $f : \mathbb{R} \to \mathbb{R}$ defined by $f(x) = x^3$ is injective.

Solution

To show that f is an injection we must prove that (3) holds. In other words, we must establish the followins implication

$$f(x) = f(y) \Rightarrow x = y . \qquad (4)$$

But $f(x) = f(y) \Rightarrow x^3 = y^3$

$$\Rightarrow x^3 - y^3 = 0$$

$$\Rightarrow (x-y)(x^2 + xy + y^2) = 0$$

$$\Rightarrow \begin{cases} x - y = 0 \\ or \\ x^2 + xy + y^2 = 0 \end{cases}$$

$$\Rightarrow \begin{cases} x = y \\ or \\ x = y = 0, \quad \text{see Problem 5 of Chapter 3} \end{cases}$$

$$\Rightarrow x = y ,$$

which establishes (4) . Therefore, f is indeed an injection; and so too the proof is complete . ♥

Example 3: Show that a function f is not injective means it satisfies the condition:

$$\exists x \in D_f, \ \exists y \in D_f, \ f(x) = f(y) \ \textit{yet} \ x \neq y . \qquad (5)$$

Solution

Since the sentence "f is not injective" is the denial of the sentence "f is injective", we have:

$$f \text{ is not injective}$$
$$\Updownarrow$$
$$\sim\left[\forall x \in D_f, \ \forall y \in D_f, \ f(x) = f(y) \rightarrow x = y\right]$$
$$\Updownarrow$$
$$\exists x \in D_f, \ \sim\left[\forall y \in D_f, \ f(x) = f(y) \rightarrow x = y\right]$$
$$\Updownarrow$$
$$\exists x \in D_f, \ \exists y \in D_f, \ \sim\left[f(x) = f(y) \rightarrow x = y\right]$$
$$\Updownarrow$$
$$\exists x \in D_f, \ \exists y \in D_f, \ \sim\left(\sim\left[f(x) = f(y)\right] \ or \ x = y\right)$$
$$\Updownarrow$$
$$\exists x \in D_f, \ \exists y \in D_f, \ f(x) = f(y) \ \textit{and} \ x \neq y ,$$

which is what we wanted to show. ♥

Example 4: Show that the function $f : \mathbb{R} \rightarrow \mathbb{R}$ defined by $f(x) = x^2$ is not injective

Solution

To show that f is not injective we may use (5). That is, we ask whether f satisfies (5). In other words, are there two distinct real numbers that share the same image by the function f ? Yes indeed, since $f(1) = f(-1)$. ♥

Exercise 5: Prove that the function $f : \mathbb{N}\times\mathbb{N} \to \mathbb{N}$ defined by $f((n,m)) = 2^m(2n-1)$ is injective.

Solution

To show that f is injective we must show that, given any two ordered pairs (a,b) and (n,m) in $\mathbb{N}\times\mathbb{N}$,

$$f((a,b)) = f((n,m)) \quad \Rightarrow \quad (a,b) = (n,m).$$

So let us set $f((a,b)) = f((n,m))$. Then, [by the Axiom of Trichotomy] we may distiguish exactly three cases:

Case 1. $b < m$:

$$f((a,b)) = f((n,m))$$

$$\Downarrow$$

$$2^b(2a-1) = 2^m(2n-1)$$

$$\Downarrow$$

$$2a-1 = 2^{m-b}(2n-1), \qquad m-b \in \mathbb{N}$$

$$\Downarrow$$

$$m-b=0,$$

since assuming otherwise would mean that a number can be at the same time odd and even.

$$\Downarrow$$

$$m=b$$

$$\Downarrow$$

CONT.

Case 2. $b > m$:

$$f((a,b)) = f((n,m))$$

$$\Downarrow$$

$$2^b(2a-1) = 2^m(2n-1)$$

$$\Downarrow$$

$$2^{b-m}(2a-1) = 2n-1, \qquad b-m \in \mathbb{N}$$

$$\Downarrow$$

$$b - m = 0,$$

since assuming otherwise would mean that a number can be at the same time odd and even.

$$\Downarrow$$

$$b = m$$

$$\Downarrow$$

$$CONT.$$

Case 2. $b = m$:

$$\begin{aligned} f((a,b)) = f((n,m)) &\Rightarrow 2^m(2a-1) = 2^m(2n-1) \\ &\Rightarrow 2^{m-m}(2a-1) = 2n-1 \\ &\Rightarrow 2a-1 = 2n-1 \\ &\Rightarrow a = n \\ &\Rightarrow (a,b) = (n,m). \end{aligned}$$

Therefore, by recapitulating the above findings, we have:

$$f((a,b)) = f((n,m))$$

$$\Downarrow$$

$$\text{CONT} \vee \text{CONT} \vee [(a,b) = (n,m)].$$

That is

$$f((a,b)) = f((n,m)) \Rightarrow (a,b) = (n,m),$$

which completes the proof. ♥

Next, we prove our first theorem. Being equivalent to Definition 1, this theorem could have been chosen as the definition of injectivity.

Theorem 6: A function $f : D_f \to C_f$ is **injective,** if and only if

$$\forall x \in D_f, \ \forall y \in D_f, \ x \neq y \to f(x) \neq f(y). \tag{6}$$

Remark 7: In other words, a function f is injective, provided no two distinct items in its domain share the same image. This is the same as saying that no item of the codomain C_f is allowed to be the image of more than one element in the domain D_f. ■

Proof [of Theorem 6]

The proof of this theorem is based on the simple fact that the two sentences

$$x \neq y \rightarrow f(x) \neq f(y)$$

and

$$f(x) = f(y) \rightarrow x = y$$

are contrapositives of one another. Indeed, since a sentence is equivalent to its contrapositive, we may substitute $f(x) = f(y) \rightarrow x = y$ for $x \neq y \rightarrow f(x) \neq f(y)$ in (3). The result is clearly (6); that is to say what we wished to obtain. ♣

15.2. SURJECTIVE[19] FUNCTIONS

We learned in Chapter 14 that every element x in the domain of a function $f : D_f \rightarrow C_f$ must be assigned to a unique element y of its codomain; and we named that element $y \in C_f$ the image of $x \in D_f$. To further help us to fix our ideas in mind, we now introduce the concept of a **preimage**.

Definition 8 [Preimage]: Let $f : D_f \rightarrow C_f$ be a function. An element x in the domain D_f is said to be a **pre-image** of an item y of the codomain C_f, if

$$f(x) = y . \qquad (7)$$

In other words, x is **a** preimage of y, simply if y is **the** image of x.

[19] A **surjective** function is sometimes called an **onto** function.

Remark 9: Note that instead of the definite article "the" we have used the indefinite article "a" in the phrase "a preimage". Indeed, the reader is reminded that an element y of the codomain C_f may have more than one pre-image. This is in contrast with the fact that every element x of the domain D_f must be, by definition of a function, associated with one and only one image $y \in C_f$. ■

We now introduce another concept, that of the **range** of a function.

Definition 10 [Range of a Function]: The **range** of a function

$$f : D_f \to C_f$$

is the set,

$$R_f = \{ y \in C_f : \ \exists x \in D_f, \ f(x) = y \}, \tag{8}$$

of all elements y in C_f that admit at least one preimage x in D_f.

Example 11: Find the range R_f of the function defined by the formula: $f(x) = \sqrt{2 - 3x}$.

Solution

As with most questions concerning functions, it will be wise here to first specify the maximal domain D_f and codomain C_f of f, even though we are not explicitly required to do so.

We know from the preceding chapter that

$$D_f = \{ x \in \mathbb{R} : \ f(x) \in \mathbb{R} \} \qquad \text{and} \qquad C_f = \mathbb{R}.$$

Thus,

$$\begin{aligned} D_f &= \{ x \in \mathbb{R} : \ \sqrt{2 - 3x} \in \mathbb{R} \} \\ &= \{ x \in \mathbb{R} : \ 2 - 3x \geq 0 \} \\ &= \left\{ x \in \mathbb{R} : \ x \leq \frac{2}{3} \right\}. \end{aligned}$$

With the domain of f found, we can now procede to determine its range by using (8):

$$\begin{aligned}
R_f &= \{y \in C_f : \exists x \in D_f \ and \ \sqrt{2-3x} = y\} \\
&= \{y \in \mathbb{R} : \exists x \leq 2/3 \ and \ \sqrt{2-3x} = y\} \\
&= \{y \geq 0 : \ \exists x \leq 2/3 \ and \ \sqrt{2-3x} = y\} \\
&= \left\{y \geq 0 : \exists x \leq 2/3 \ and \ x = \tfrac{2-y^2}{3}\right\} \\
&= \left\{y \geq 0 : \tfrac{2-y^2}{3} \leq \tfrac{2}{3}\right\} \\
&= \left\{y \geq 0 : -y^2 \leq 0\right\} \\
&= \left\{y \in \mathbb{R} : \quad y \geq 0 \ and \ -y^2 \leq 0\right\} \\
&= \left\{y \in \mathbb{R} : \quad y \geq 0\right\} \cap \left\{y \in \mathbb{R} : \quad -y^2 \leq 0\right\} \\
&= [0, \ +\infty[\ \cap \mathbb{R} \\
&= [0, \ +\infty[\ . \ \heartsuit
\end{aligned}$$

Remark 12: From the foregoing example, we can easily see that the range of the function $f : (-\infty, 2/3] \to \mathbb{R}$ defined by the formula

$$f(x) = \sqrt{2-3x}$$

is not the entire codomain $\mathbb{R}$. ■

Definition 13 [Surjection]: A function $f : D_f \to C_f$ is said to be **surjective** or **a surjection**, if its range R_f is its entire codomain Y :

$$R_f = C_f$$

Remark 14: We know that $R_f = C_f$ means $R_f \subset C_f$ and $C_f \subset R_f$. We also know that the first inclusions, $R_f \subset Y$, is always true by definition of the range. Therefore, Definition 13 may be rephrased as follows: *A function* $f : D_f \to C_f$ *is a surjection, if* $C_f \subset R_f$. In other words, a function $f : D_f \to C_f$ is a surjection, if every element of its codomain C_f admits at least one preimage in its domain D_f. ■

Example 15: Why is the function $f :]-\infty, 2/3] \to \mathbb{R}$ defined by the formula $f(x) = \sqrt{2 - 3x}$ not a surjection?

Solution

This function is not a surjection, since not every element of its codomain, $\mathbb{R}$, has a pre-image. For instance, the number -2 has no pre-image; in fact, there is clearly no element in the domain $]-\infty, 2/3]$ whose image is negative. To get a surjective function out of the present function, we may reduce the codomain, $\mathbb{R}$, to the range $[0, +\infty[$. But that will result in a different function, $f :]-\infty, 2/3] \to [0, +\infty[$, from the one we started. ♥

Example 16: Show that the function

$$f :]-\infty, -1[\cup]-1, 1[\cup]1, +\infty[\to \mathbb{R}$$

defined by the formula $f(x) = \dfrac{x}{x^2 - 1}$ is a surjection?

Solution

To show that f is a surjection, we must show that every number y in the codomain $\mathbb{R}$ has at least one pre-image x in $]-\infty, -1[\cup]-1, 1[\cup]1, +\infty[$.

So, let $y \in \mathbb{R}$ and find an x in $]-\infty, -1[\cup]-1, 1[\cup]1, +\infty[$ such that

$$f(x) = y. \tag{9}$$

To do this, we will consider two cases:

Case 1. $y = 0$: In this case, all we have to do is to choose $x = 0$, and (9) will then be satisfied.

Case 2. $y \neq 0$: Then we have

$$f(x) = y \iff \frac{x}{x^2 - 1} = y$$

$$\iff yx^2 - x - y = 0$$

$$\iff \begin{cases} x = \dfrac{1 + \sqrt{1 + 4y^2}}{2y} \\ or \\ x = \dfrac{1 - \sqrt{1 + 4y^2}}{2y} \end{cases}$$

Hence, by choosing

$$x = \frac{1 + \sqrt{1 + 4y^2}}{2y},$$

we get not just $f(x) = y$ but also $x \in D_f$. Therefore, the proof is now complete. ♥

15.3. BIJECTIVE FUNCTIONS

In the preceding sections we introduced the ideas of injectivity and surjectivity. This was, in part, in preparation for the present section. As will be seen in the current section, functions that are both injective and surjective enjoy many interesting properties.

But before going any further, a remark on our notations is in order: In the remainder of this chapter we will denote the domain D_f of an arbitrary function f simply by X, and the codomain C_f by Y. This is in an attempt to bring notational simplification where that may prove advantageous.

Definition 17 [Bijection]: A function $f : X \to Y$ is said to be **bijective** or **a bijection,** if it is both an injection and a surjection.

Remark 18: Thus, in light of Remarks 7 and 14, a function $f : X \to Y$ is a bijection, if every element of its codomain Y admits **one and only one** pre-image in its domain X. ■

Example 19
Let $\mathbb{E}$ be the set of even [natural] numbers, and $\mathbb{O}$ the set of odd [natural] numbers. We are asking that you find a bijective function f defined on the set $\mathbb{E}$ onto the set $\mathbb{O}$.

Solution
By letting

$$f : \mathbb{E} \to \mathbb{O} \quad \text{such that} \quad f(n) = n - 1,$$

we can easily see that f is a bijection defined on $\mathbb{E}$ with values in the set $\mathbb{O}$. ♥

As a direct consequence of the foregoing remark we have:

Theorem 20: Assume $f : X \to Y$ is a bijection. Then there is a function $g : Y \to X$ such that

$$g o f = I_X \quad \text{and} \quad f o g = I_Y, \tag{10}$$

where I_X and I_Y are the identity functions on set X and set Y, respectively.

Proof:
Let $f : X \to Y$ be a bijection. Then, by Definition 17, f is a surjection, which means that every $y \in Y$ admits a pre-image x in X. Furthermore, we know that f is also an injection. Hence, the preimage x associated with each $y \in Y$ is unique.

Hence, let us denote by g the rule which assigns to every element $y \in Y$ its unique preimage x in X. Clearly, the function $g : Y \to X$ as constructed satisfies:

$$g(y) = x \quad \textit{if and only if} \quad f(x) = y \,. \tag{11}$$

Therefore, to complete the proof, we just need to show that both relations in (10) hold true. But these are direct consequences of (11). Indeed, from (11), we have:

$$\forall x \in X, \quad g\left[f(x)\right] = g(y) = x$$

and

$$\forall y \in Y, \quad f\left[g(y)\right] = f(x) = y \,.$$

That is,

$$\forall x \in X, \quad g\left[f(x)\right] = x \qquad \text{and} \qquad \forall y \in Y, \quad f\left[g(y)\right] = y \,,$$

which establishes both relations in (10). ♣

Definition 21 [Inverse function]: Let $f : X \to Y$ be a bijection. Then the function $g : Y \to X$ with properties as stated in Theorem 20 is called the **inverse function** of f and is denoted by f^{-1}.

Remark 22: Clearly, the phrase **inverse function** is motivated by the fact that we have:

$$\forall x \in X, \; f^{-1}\left[f(x)\right] = x \; ; \text{and} \; \forall y \in Y, \; f\left[f^{-1}(y)\right] = y \,.$$

In other words, what the function f does to x, the function f^{-1} undoes [or reverses] it, and vice versa. ■

Next, we show that a bijection can have only one inverse

Theorem 23: The inverse f^{-1} of a bijection f is unique.

Proof by contradiction:

Let $f : X \to Y$ be a bijection. Now, assume contrary to the theorem that f has at least two distinct inverses:

$$g : Y \to X \quad \text{and} \quad h : Y \to X .$$

We will show that this assumption leads to a contradiction. To do so, let us study the composite function $gofoh$. We know from Problem 5 of Chapter 14 that:

$$(gof)oh = gofoh = go(foh),$$

from which we deduce

$$I_X oh = gofoh = goI_Y,$$

that is

$$h = gofoh = g .$$

Thus,

$$h = g,$$

which contradicts the assumption made earlier to suggesting that the functions g and h are distinct. ♣

Theorem 24: Let there be two functions $f : X \to Y$, and $g : Y \to X$ such that $gof = I_X$. Then f is injective, and g is bijective.

Proof:

- To show that f is injective, we must show that for every x and y in X, if $f(x) = f(y)$ then $x = y$.

  Clearly, $f(x) = f(y) \Rightarrow g[f(x)] = g[f(y)]$, since an image by g is unique.

  $$\Rightarrow [gof](x) = [gof](y)$$

  $$\Rightarrow I_X(x) = I_X(y)$$

  $$\Rightarrow x = y .$$

  Thus, f is indeed injective.

- As for the surjectivity of g, we must establish that every element x of X admits a preimage y in Y by the function g.

  Since by assumption $gof = I_X$, we have, for every x in X,

  $$[gof](x) = I_X(x),$$

  that is

  $$g[f(x)] = x,$$

  or

  $$g(y) = x \quad \text{where we have set} \quad y = f(x) \in Y.$$

  This shows that $y \in Y$ is a preimage of $x \in X$. Therefore, function g is surjective. ♣

Thus, we arrive at the following more comprehensive result:

Corollary 25: Every bijection $f : X \to Y$ has a unique inverse $f^{-1} : Y \to X$ which is also a bijection

Proof:
This corollary is an easy consequence of Theorems 20, and 24 . ♣

15.4. IMAGES AND INVERSE IMAGES OF SETS

We are now going to introduce two more concepts that will be needed in later chapters. These are the concept of the **image of a set** and that of the **inverse image of a set**.

Definition 26: Let a function $f : X \to Y$ and a subset A of its domain X be given. Then the set,

$$f(A) = \{ y \in Y : \quad \exists x \in A, \quad f(x) = y \}, \tag{12}$$

of all elements of Y that have preimages in A is called the **image of** A **under** f.

Remark 27: Thus, it is easy to see that

$$R_f = \{ y \in Y : \quad \exists x \in X, \ \ f(x) = y \} = f(X).$$

That is, the range of a function $f : X \to Y$ is the image of the domain of the function . ■

Exercise 28 Let $f : X \to Y$ be a function, and A and B subsets of X. Show that

1. $f(A \cup B) = f(A) \cup f(B)$.
2. $f(\varnothing) = \varnothing$.

Solution

1.
$$\begin{aligned} f(A \cup B) &= \{ y : \ \exists x \in A \cup B, \ f(x) = y \} \\ &= \{ y : \ \exists x \in A \cup B \\ &\qquad\qquad and \ \ f(x) = y \} \\ &= \{ y : \ (\exists x \in A \ \ or \ \exists x \in B) \\ &\qquad\qquad and \ \ (f(x) = y) \} \\ &= \{ y : \ (\exists x \in A \ \ and \ \ f(x) = y) \\ &\qquad\qquad or \ \ (\exists x \in B \ \ and \ \ f(x) = y) \} \\ &= \{ y : \ \exists x \in A, \ f(x) = y \} \\ &\qquad \cup \{ y : \ \exists x \in B, \ f(x) = y \} \\ &= f(A) \cup f(B). \end{aligned}$$

2. We will prove this part of the theorem by **contradiction**. Assume $f(\varnothing) = \varnothing$ is false. This means that $f(\varnothing)$ is nonempty. Hence, there is at least one item, say y, in $f(\varnothing)$. Therefore, by Definition 26,

$$\exists x \in \varnothing, \ \ f(x) = y ,$$

which is impossible; for indeed, $\varnothing$ contains no elements, let alone an object x. ♥

Definition 29: Let a function $f: X \to Y$ and a subset B of its codomain Y be given. Then the set,

$$f^{-1}(B) = \{ x \in X : \quad f(x) \in B \} , \tag{13}$$

of all elements of the domain X that have their images in B is called the **inverse image of B under the function f .**

Remark 30: Thus, it is obvious that the domain X of any function $f: X \to Y$ is the inverse image of its codomain Y . ■

Exercise 31

Let $f: X \to Y$ be a function, and A and B subsets of Y. Show that

1. $f^{-1}(A \cup B) = f^{-1}(A) \cup f^{-1}(B)$.
2. $f^{-1}(A \cap B) = f^{-1}(A) \cap f^{-1}(B)$.
3. $f^{-1}(\varnothing) = \varnothing$.

Solution

1.
$$\begin{aligned} f^{-1}(A \cup B) &= \{ x : \quad f(x) \in A \cup B \} \\ &= \{ x : \ (f(x) \in A) \ or \ (f(x) \in B) \} \\ &= \{ x : \ f(x) \in A \} \\ &\qquad \cup \{ x : \ f(x) \in B \} \\ &= f^{-1}(A) \cup f^{-1}(B). \end{aligned}$$

2.
$$f^{-1}(A \cap B) = \{ x : \quad f(x) \in A \cap B \}$$

$$= \{x : f(x) \in A \text{ and } f(x) \in B\}$$

$$= \{x : f(x) \in A\} \cap \{x : f(x) \in B\}$$

$$= f^{-1}(A) \cap f^{-1}(B).$$

3. We will prove this part of the theorem by contradiction. Assume $f^{-1}(\varnothing) = \varnothing$ is false. This means that $f^{-1}(\varnothing)$ is nonempty. Hence, there is at least one item, say x, in $f^{-1}(\varnothing)$. Therefore, by Definition 29,

$$f(x) \in \varnothing,$$

which is impossible; for indeed, $\varnothing$ contains no elements, let alone the object $f(x)$. ♥

PROBLEMS

1. In each case below find the range of the function.

   1.1. $f(x) = \sqrt{12 - 3x}$.

   1.1. $f(x) = \dfrac{1}{3 + x}$.

   1.1. $f(x) = x^2 - x + 1$.

2. Let $f : \mathbb{N} \times \mathbb{N} \to \mathbb{N}$ be the function defined by:

$$f(n, m) = 2^n \cdot 3^m, \text{ for } (n, m) \in \mathbb{N} \times \mathbb{N}.$$

whih ones of these properties does f have?

   2.1. Injectivity.

2.2. Surjectivity.

2.3. Bijectivity.

3. Prove that if

$$u : X \to Y, \quad \textit{and} \quad v : Y \to Z$$

are both bijections, their composition

$$vou : X \to Z$$

is also a bijection. [Hint: First show that $(u^{-1}ov^{-1})o(vou) = I_X$ and $(vou)o(u^{-1}ov^{-1}) = I_Z$.]

4. * Prove the following principle called the **Gluing Principle**:

Let both

$$u : A_1 \to B_1,$$

and

$$v : A_2 \to B_2$$

be bijections, where $A_1 \cap A_2 = \varnothing = B_1 \cap B_2$. Then the function

$$f : A_1 \cup A_2 \to B_1 \cup B_2,$$

defined by

$$f(x) = \begin{cases} u(x), & \textit{if} \quad x \in A_1 \\ v(x), & \textit{if} \quad x \in A_2 \end{cases}$$

is also a bijection.

5. Let $f : X \to Y$ be a function, and A a subset of X, and B a subset of Y. Then show that

1. $f^{-1}(Y - B) = X - f^{-1}(B)$.
2. $f(X - A) = Y - f(A)$ is not true in general.

6. In the Calculus, a function $f:]a, b[\to \mathbb{R}$ is said to be increasing if:

$$\textit{whenever } x < y \textit{ then so is } f(x) < f(y).$$

Using Theorem 6, the Axiom of Trichotomy, and the method of proof by exhaustion of all cases, prove that an increasing function is injective.

7. * Show that any identity function $I: S \to S$ is bijective.

"All human knowledge thus begins with intuitions, proceeds thence to concepts, and ends with ideas."
David Hilbert.

CHAPTER 16

SEQUENCES

In Chapter 14 we introduced the general idea of a function; in this chapter we will study a special class of functions called **sequences**. A sound understanding of the theory of sequences is useful to anyone planning to study advanced mathematical subjects such as the *Calculus* and *Topology*.

16.1. DEFINITIONS OF A SEQUENCE

The purpose of this section is to define the concept of a sequence.

Definition 1 [Sequences]: A **sequence** in mathematics is any function, s, whose domain is the set of natural numbers $\mathbb{N}$. [As for its codomain, it may be any set C_s.]

$$s : \mathbb{N} \to C_s .$$

Example 2:
Explain why these functions are sequences?

1. $s : \mathbb{N} \to [0, \infty[$, defined by the formula $s(n) = ar^n$, where $a > 0$ and $r > 0$.

2. $p : \mathbb{N} \to \mathbb{R}$, defined by the formula $p(n) = a + nd$, where a and d are in $\mathbb{R}$.

3. $f : \mathbb{N} \to \mathbb{Q}$, defined by $f(n) = \dfrac{2n+4}{n^2+1}$.

4. $h : \mathbb{N} \to \mathbb{N}$, defined by $h(n) = 2n^2 + 1$.

Solution

All of these functions qualify as sequences, since they all have the set $\mathbb{N}$ as their domain. ♥

Definition 3:

1. A sequence $s : \mathbb{N} \to C_s$ is called a **sequence of real numbers**, if $C_s \subset \mathbb{R}$.

3. A sequence $s : \mathbb{N} \to C_s$ is called a **sequence of rational numbers**, if $C_s \subset \mathbb{Q}$.

2. A sequence $s : \mathbb{N} \to C_s$ is called a **sequence of integers**, if $C_s \subset \mathbb{Z}$.

3. A sequence $s : \mathbb{N} \to C_s$ is called a **sequence of natural numbers**, if $C_s \subset \mathbb{N}$.

Example 4:

What types of sequences are the sequences given in Example 2 above?

Solution

1. Sequence of real numbers.

2. Sequence of real numbers.

3. Sequence of rational numbers.

4. Sequence of natural numbers.

Note that a sequence may fall into more than one type. For example, sequence 4 is also a sequence of integers, a sequence of rational numbers and a sequence of real numbers, since

$$C_s \subset \mathbb{N} \subset \mathbb{Z} \subset \mathbb{Q} \subset \mathbb{R}.$$

Hence, you may have noticed that, in a case such as this, we tend to name the sequence after the smallest set. But this practice of retaining the smallest set is totally arbitrary. ♥

Remark 5: Given any sequence $s : \mathbb{N} \to C_s$, note that by letting

$$s(1) = s_1, \quad s(2) = s_2, \quad s(3) = s_3, \quad s(4) = s_4, \ . \ . \ . \qquad (1)$$

we obtain an infinite list of items

$$s_1, \quad s_2, \quad s_3, \quad s_4, \ . \ . \ . \qquad (2)$$

in C_s. And conversely, from any infinite list of objects (2) belonging to a set C_s, we may define a sequence, $s : \mathbb{N} \to C_s$ via (1). ■

Thus, Definition 1 is equivalent to the following

Definition 6 [Sequences]: By a sequence of elements of a set C_s, we mean any infinite list

$$s_1, \quad s_2, \quad s_3, \quad s_4, \ . \ . \ . \qquad (3)$$

of elements belongings to set C_s . For notational convenience, the sequence (3) is often time written more compactly as $(s_n)_{n=1}^{\infty}$.

Remark 7: Sometimes, a sequence may be described by a rule that expresses each element of the sequence in terms of earlier elements. **Such a sequence is said to be defined recursively**. For example, the following sequences are given recursively:

1. $(s_n)_{n=1}^{\infty}$ where $s_1 = \sqrt{2}$, and $s_n = \sqrt{2 + \sqrt{s_{n-1}}}$ for any $n \geq 2$.

2. $(s_n)_{n=1}^{\infty}$ where $s_1 = 1$, $s_2 = 1$, and $s_n = s_{n-1} + s_{n-2}$ for any $n \geq 3$. ■

In the remainder of this chapter we will assume that all our sequences are sequences of real numbers. Having made that supposition, we will now proceed to group all sequences into two distinct categories: **convergent sequences** and **divergent sequences**. We will start with convergent sequences.

16.2. CONVERGENCE AND DIVERGENCE

What is a **convergent sequence**? To better answer this question, we begin with the following illustration. Image that we have of a piece of land 6 square miles large and we decide to donate 1 square mile [to a local college], then the following year half of a square mile from what is left, then the year after a quarter of a square mile from the new remainder, and so on indefinitely. Then clearly, the areas of the remainders form the unending sequence:

$$6\,\text{mi}^2, \quad 5\,\text{mi}^2, \quad 4\tfrac{1}{2}\,\text{mi}^2, \quad 4\tfrac{1}{4}\,\text{mi}^2, \quad 4\tfrac{1}{8}\,\text{mi}^2, \quad 4\tfrac{1}{16}\,\text{mi}^2, \ldots$$

Note that, not only is there always more than 4 squre miles of land left, but the remainder of land may be made as close to 4 square miles as we wish by simply continuing the donation process long enough. Thus 4, which is the mumber steadily approached by the remainders is called the limit of the sequence.

In general, assume $(s_n)_{n=1}^{\infty}$ is a sequence of real numbers; if the elements [or terms] of the sequence eventually become and remain as close as we like to some real number L , then L is called the limit of the sequence. And in that case the sequence $(s_n)_{n=1}^{\infty}$ is said to be convergent.

We are now in the position to give a more precise definition of a convergent sequence.

Definition 8 [Convergent sequences]: A sequence $(s_n)_{n=1}^{\infty}$ is said to **converge** or to **be convergent** to a real number L, if

$$\forall \varepsilon > 0, \quad \exists N_\varepsilon \in \mathbb{N}, \quad \forall n \in \mathbb{N}, \ n > N_\varepsilon \rightarrow s_n \in]L - \varepsilon, \ L + \varepsilon[. \qquad (4)$$

The number L is called the limit of the sequence $(s_n)_{n=1}^{\infty}$.

We often abbreviate the whole sentence (4) as

$$\lim_{n \to \infty} s_n = L \qquad \text{or as} \qquad s_n \to L \ \textit{when} \ n \to \infty , \qquad (5)$$

and say that the limit of s_n is L when n goes to infinity; or that s_n approaches to L as n goes to infinity.

Remark 9: The foregoing definition in symbolic form may seem confusing at first. But, after some rereading, what it says is pretty simple to undertand. It says, given any open interval $]L - \varepsilon, \ L + \varepsilon[$ centered on L, no matter how narrow, there is a corresponding natural number, expressed as N_ε depending on ε, such that every natural number n greater than N_ε has its image s_n in $]L - \varepsilon, \ L + \varepsilon[$. ■

Now, from the result of Problem 4 of Chapter 14, we do know that $s_n \in]L - \varepsilon, \ L + \varepsilon[\ \Leftrightarrow \ | s_n - L | < \varepsilon$. Therefore, we can rewrite Definition 8 above as follows

Definition 10 [Convergent sequence]: A sequence $(s_n)_{n=1}^{\infty}$ is said to **converge** or to **be convergent** to a real number L, if

$$\forall \varepsilon > 0, \quad \exists N_\varepsilon \in \mathbb{N}, \quad \forall n \in \mathbb{N}, \ n > N_\varepsilon \rightarrow | s_n - L | < \varepsilon . \qquad (6)$$

Hence, in plain English, we say that a sequence $(s_n)_{n=1}^{\infty}$ converges towards L, if for any real number ε, no matter how small, there exists a **term** of the sequence, say s_{N_ε}, all of whose successors each differ from L by less than ε.

Example 11: Show that the sequence $\left(\frac{1}{n}\right)_{n=1}^{\infty}$ is convergent.

Solution

Intuitively, we suspect that the sequence $\left(\frac{1}{n}\right)_{n=1}^{\infty}$ converges to 0. But we cannot be certain until and unless we are able to establish this rigorously. To prove the convergence of $\left(\frac{1}{n}\right)_{n=1}^{\infty}$ to 0, we must show that

$$\forall \varepsilon > 0, \quad \exists N \in \mathbb{N}, \quad \forall n \in \mathbb{N}, \; n > N \to \left| \tfrac{1}{n} - 0 \right| < \varepsilon,$$

which, once simplified, is the same as

$$\forall \varepsilon > 0, \quad \exists N \in \mathbb{N}, \quad \forall n \in \mathbb{N}, \; n > N \to \tfrac{1}{n} < \varepsilon. \tag{7}$$

To show that (7) holds, let us be given **any** $\varepsilon > 0$. Then we know, from Exercise 31.2 of Chapter 11, that **there is** a natural number N such that

$$\tfrac{1}{N} < \varepsilon. \tag{8}$$

We also know, from Theorem 39.5 of Chapter 9, that

$$\forall n \in \mathbb{N}, \; n > N \to \tfrac{1}{n} < \tfrac{1}{N}. \tag{9}$$

Thus, combining (8) and (9), we conclude that

$$\forall n \in \mathbb{N}, \; n > N \to \tfrac{1}{n} < \tfrac{1}{N} < \varepsilon, \tag{10}$$

which establishes (7) . ♥

Remark 12:

It is important to keep in mind that the choice of N is in general dependent upon the value of ε. For instance, in Example 11, the smaller we pick ε, the larger N will be. Thus to be more rigorous in our notation, we should have written N_ε in place of N. But if we sometimes write simply N , it is because we wish not to complicate our notations. ■

Example 13: Show that these sequences are convergent.

1. $(c)_{n=1}^{\infty}$, where c is a fixed number of $\mathbb{R}$.

2. $\left(\frac{A}{n^{\alpha}}\right)_{n=1}^{\infty}$, where A is a fixed number of $\mathbb{R}$, and $1 < \alpha$.

3. $(c^n)_{n=1}^{\infty}$, where $0 < c < 1$.

Solution

1. The convergence of $(c)_{n=1}^{\infty}$ to c is shown by simply verifying that $(c)_{n=1}^{\infty}$ satisfies the criterion (4) or (6). Since this is not at all difficult, we leave it to the reader to supply all the details.
2. We can show that $\left(\frac{A}{n^{\alpha}}\right)_{n=1}^{\infty}$ is convergent to 0 by using the result established in Example 11 and the fact that $\frac{1}{n^{\alpha}} \leq \frac{1}{n}$, for all $n \geq 1$. Since this is not at all difficult, we leave it to the reader to supply all the details

3. The arguments for the convergence of $(c^n)_{n=1}^{\infty}$ to 0 are similar to those used in Example 11. You must however make use of the exponential version of the Archimedean principle instead; see Exercise 33.2 of Chapter 11. We leave it to the reader to supply all the details, since this is easily done. ♥

With the foregoing examples we realize that there is indeed such a thing as a convergent sequence. But, as will be seen next, not all sequences are convergent.

Definition 14: A sequence $(s_n)_{n=1}^{\infty}$ is said to be **divergent**, if it is not convergent [to any real number.]

Exercise 15: Show that a divergent sequence $(s_n)_{n=1}^{\infty}$ is a sequence that satisfies the condition:

$$\begin{aligned} &\forall L \in \mathbb{R},\ \exists \varepsilon_L > 0,\ \forall N \in \mathbb{N}, \\ &\exists n_N \in \mathbb{N},\ n_N > N \ \textit{and} \ | s_{n_N} - L | \geq \varepsilon. \end{aligned} \tag{11}$$

Solution

First of all, using Definition 8, let us notice that $(s_n)_{n=1}^{\infty}$ would be convergent, if and only if

$$\exists L \in \mathbb{R},\ \forall \varepsilon > 0,\ \exists N \in \mathbb{N},\ \forall n \in \mathbb{N},\ n > N \rightarrow | s_n - L | < \varepsilon.$$

Therefore, taking the denial of the above, $(s_n)_{n=1}^{\infty}$ becomes divergent, if

$$\sim \left(\exists L \in \mathbb{R},\ \forall \varepsilon > 0,\ \exists N \in \mathbb{N},\ \forall n \in \mathbb{N},\ n > N \rightarrow | s_n - L | < \varepsilon\right),$$

which is the same as

$$\begin{aligned} &\forall L \in \mathbb{R},\ \exists \varepsilon_L > 0,\ \forall N \in \mathbb{N}, \\ &\exists n_N \in \mathbb{N},\ \sim\left(n_N > N \rightarrow | s_{n_N} - L | < \varepsilon_L\right), \end{aligned}$$

which in turn is the same as

$$\begin{aligned} &\forall L \in \mathbb{R},\ \exists \varepsilon_L > 0,\ \forall N \in \mathbb{N}, \\ &\exists n_N \in \mathbb{N},\quad n_N > N \ \textit{and yet} \ | s_{n_N} - L | \geq \varepsilon_L \ . \end{aligned}$$ ♥

Example 16: Prove that the sequence $\left([-1]^n\right)_{n=1}^{\infty}$ is divergent.

Solution

To prove the divergence of $\left([-1]^n\right)_{n=1}^{\infty}$, we must show that

$$\forall L \in \mathbb{R},\ \exists \varepsilon_L > 0,\ \forall N \in \mathbb{N},$$
$$\exists n_N \in \mathbb{N},\quad n_N > N \ \ and\ \ |\, s_{n_N} - L \,| \geq \varepsilon\ ,$$

which is the same as

$$\begin{gathered}\forall L \in \mathbb{R},\ \exists \varepsilon_L > 0,\ \forall N \in \mathbb{N},\\ \exists n_N \in \mathbb{N},\quad n_N > N \ \ and\ \ s_n \notin \,]\, L - \varepsilon_L,\ L + \varepsilon_L\, [\ .\end{gathered} \tag{12}$$

To show that (12) holds for this particular sequence, let us be given **any** $L \in \mathbb{R}$ **. Then we may distinguish three cases**:

Case 1. $L < 1$:
Here, by letting $\varepsilon = 1 - L$, we have $]L - \varepsilon,\ L + \varepsilon[\, = \,]2L - 1,\ 1[$. Hence, for any $N \in \mathbb{N}$, by picking $n = 2N$, we see that $n > N$ and $s_n = [-1]^n = 1 \notin \,]L - \varepsilon,\ L + \varepsilon[$, which is what we wanted to show.

Case 2. $L = 1$:
In this case, by choosing for example $\varepsilon = 1$, we get $]L - \varepsilon,\ L + \varepsilon[=]0,\ 2[$. Thus, for any $N \in \mathbb{N}$, by picking $n = 2N + 1$, we have at the same time $n > N$ and $s_n = [-1]^n = -1 \notin]L - \varepsilon,\ L + \varepsilon[$, which is what we sought to prove in this case too.

Case 3. $L > 1$:
This case is similar to the first case. Indeed, in this case, by letting $\varepsilon = L - 1$, we have $]L - \varepsilon,\ L + \varepsilon[=]1,\ 2L - 1[$. Hence, for any $N \in \mathbb{N}$, by picking $n = 2N$, we again get both $s_n = [-1]^n = 1 \notin]L - \varepsilon,\ L + \varepsilon[$ and $n > N$, which is what we wished to show.

Thus the entire proof is now complete. ♥

16.3. TWO IMPORTANT PROPERTIES OF CONVERGENT SEQUENCES

Now that we know what a convergent sequence is, we will devote this section to studying some of the properties of this type of sequences. These properties are given here as theorems, the first of which is

Theorem 17 [Uniqueness of the limit]: A convergent sequence $(s_n)_{n=1}^{\infty}$ has one, and only one, limit L.

Proof by contradiction

Suppose $(s_n)_{n=1}^{\infty}$ is a convergent sequence **with at least two distinct limit L_1 and L_2**. We will show that this assumption leads to a contradiction.

To say that L_1 and L_2 are limits of $(s_n)_{n=1}^{\infty}$ means that **for every** $\varepsilon > 0$,

$$\exists N_1 \in \mathbb{N},\ \ \forall n \in \mathbb{N},\ \ n > N_1 \to | s_n - L_1 | < \tfrac{\varepsilon}{2}. \tag{13}$$

and

$$\exists N_2 \in \mathbb{N},\ \ \forall n \in \mathbb{N},\ \ n > N_2 \to | s_n - L_2 | < \tfrac{\varepsilon}{2}. \tag{14}$$

Now, let us denote by N the larger of the two natural numbers N_1 and N_2. Then,

$$\forall n \in \mathbb{N},\ \ n > N \to | s_n - L_1 | < \tfrac{\varepsilon}{2} \tag{15}$$

and

$$\forall n \in \mathbb{N},\ \ n > N \to | s_n - L_2 | < \tfrac{\varepsilon}{2}. \tag{16}$$

Hence, by combining (15) and (16), we have:

$$\forall n \in \mathbb{N},\ \ n > N \to | L_1 - s_n | + | s_n - L_2 | < \tfrac{\varepsilon}{2} + \tfrac{\varepsilon}{2} = \varepsilon,$$

which, by virtue of the triangle inequality

$$| L_1 - L_2 | \leq | L_1 - s_n | + | s_n - L_2 |,$$

implies that

$$|L_1 - L_2| < \varepsilon \quad \textit{for every } \varepsilon > 0 .$$

Thus, $L_1 - L_2 = 0$[20]. Therefore $L_1 = L_2$, which is clearly contradictory with our assumption that the two reals L_1 and L_2 were distinct . ♣

Definition 18 : A sequence $(s_n)_{n=1}^{\infty}$ is said to be **bounded**, if there exist two **fixed** real numbers A and B such that the inequality

$$A \leq s_n \leq B,$$

holds for all $n \in \mathbb{N}$.

Exercise 19: Show that a sequence $(s_n)_{n=1}^{\infty}$ is **bounded**, if and only if there exists a **fixed** real number M such that

$$|s_n| \leq M, \quad n \in \mathbb{N}.$$

Solution

To solve this exercise, we will procede by equivalences:

$$(s_n)_{n=1}^{\infty} \text{ is } \textbf{bounded}$$

$$\Updownarrow$$

$$\exists A \in \mathbb{R},\ \exists B \in \mathbb{R},\quad \forall n \in \mathbb{N},\ A \leq s_n \leq B$$

$$\Updownarrow$$

$$\exists A \in \mathbb{R},\ \exists B \in \mathbb{R},\quad \forall n \in \mathbb{N},\ s_n \leq B \ \textit{ and } \ -s_n \leq -A$$

$$\Updownarrow$$

$$\forall n \in \mathbb{N},\ s_n \leq M \ \textit{ and } \ -s_n \leq M,$$

where M is the bigger of the two numbers $-A$ and B.

[20] Indeed, any quantity A whose absolute value is less than any positive number ε whatsoever must equal 0 .

$$\Updownarrow$$

$$\exists M \in \mathbb{R}, \quad \forall n \in \mathbb{N}, \quad -M \leq s_n \leq M$$

$$\Updownarrow$$

$$\exists M \in \mathbb{R}, \quad \forall n \in \mathbb{N}, \quad |\, s_n \,| \leq M,$$

as we wished to prove. ♥

With the above definition, we can now present our second important property of convergent sequences. This is a property that will be useful in establishing Theorem 24 of the next section.

Theorem 20: Every convergent sequence $(s_n)_{n=1}^{\infty}$ is bounded.

Proof

Let the sequence $(s_n)_{n=1}^{\infty}$ be convergent. Then,

$$\forall \varepsilon > 0, \quad \exists N \in \mathbb{N}, \quad \forall n \in \mathbb{N}, \; n > N \to |\, s_n - s \,| < \varepsilon,$$

where s is the limit of $(s_n)_{n=1}^{\infty}$. Therefore, for the special case $\varepsilon = 1$,

$$\exists N \in \mathbb{N}, \quad \forall n \in \mathbb{N}, \; n > N \to |\, s_n - s \,| < 1.$$

Hence, for every natural number $n > N$,

$$-1 < s_n - s < 1,$$

which in turn means that

$$s - 1 < s_n < s + 1, \quad \text{for all} \quad n > N.$$

Now pick the number B to be equal to the biggest among these $N+1$ numbers: $s_1, s_2, s_3, \ldots, s_N$, and $s+1$. And also choose A as the smallest of these $N+1$ numbers: $s_1, s_2, s_3, \ldots, s_N$, and $s-1$. Then, from the last inequality above, we have:

$$A \leq s_n \leq B, \quad \text{for all} \quad n \in \mathbb{N},$$

which completes the proof. ♣

16.5. OPERATIONS ON SEQUENCES

Theorem 21 [The Sum Rule]: If $(r_n)_{n=1}^{\infty}$ converges to r and $(s_n)_{n=1}^{\infty}$ converges to s, then the **sum sequence** $(r_n + s_n)_{n=1}^{\infty}$ also converges but to $r + s$.

Remark 22: In other words,

$$\lim_{n\to\infty}(r_n + s_n) = \lim_{n\to\infty} r_n + \lim_{n\to\infty} s_n \ , \tag{17}$$

provided both $\lim_{n\to\infty} r_n$ and $\lim_{n\to\infty} s_n$ exist. ■

Proof of Theorem 21

For every $\varepsilon > 0$, since $(r_n)_{n=1}^{\infty}$ and $(s_n)_{n=1}^{\infty}$ converge to r and s, respectively, we have:

$$\exists N_1 \in \mathbb{N},\ \forall n \in \mathbb{N},\ n > N_1 \to | r_n - r | < \tfrac{\varepsilon}{2}$$

and

$$\exists N_2 \in \mathbb{N},\ \forall n \in \mathbb{N},\ n > N_2 \to | s_n - s | < \tfrac{\varepsilon}{2}.$$

Therefore, denoting by N the larger of the two naturals N_1 and N_2, we obtain

$$\forall n \in \mathbb{N},\ n > N \to | r_n - r | + | s_n - s | < \tfrac{\varepsilon}{2} + \tfrac{\varepsilon}{2} = \varepsilon\ ,$$

which, thanks to the triangle inequality

$$| (r_n - r) + (s_n - s) | \le | r_n - r | + | s_n - s |\ ,$$

yields

$$\forall n \in \mathbb{N},\ n > N \to | (r_n - r) + (s_n - s) | \le | r_n - r | + | s_n - s | < \varepsilon\ .$$

This in turn implies that

$$\forall n \in \mathbb{N},\ n > N \to | (r_n + s_n) - (r + s) | = | (r_n - r) + (s_n - s) | < \varepsilon\ .$$

Hence, summing all up, we have arrive at the fact that

$$\forall \varepsilon > 0,\ \exists N \in \mathbb{N},\ \ \forall n \in \mathbb{N},\ n > N \rightarrow |\,(r_n + s_n) - (r + s)\,| < \varepsilon\,,$$

which completes the proof. ♣

Example 23: Show that the sequence

$$\left(\frac{5 + 3n^2}{6n^2}\right)_{n=1}^{\infty}$$

is convergent to $\frac{1}{2}$.

Solution

To see why this sequence is convergent to $\frac{1}{2}$, we only need to notice these three facts:

$$\frac{5 + 3n^2}{6n^2} = \frac{5/6}{n^2} + \frac{1}{2}, \qquad \lim_{n\to\infty} \frac{5/6}{n^2} = 0, \quad \text{and} \quad \lim_{n\to\infty} \frac{1}{2} = \frac{1}{2}$$

For indeed, applying the preceding theorem, we get

$$\lim_{n\to\infty} \frac{5 + 3n^2}{6n^2} = \lim_{n\to\infty} \frac{5/6}{n^2} + \lim_{n\to\infty} \frac{1}{2} = 0 + \frac{1}{2} = \frac{1}{2}.$$ ♥

Theorem 24 [The Product Rule]: If $(r_n)_{n=1}^{\infty}$ converges to r and $(s_n)_{n=1}^{\infty}$ converges to s, then the **product sequence** $(r_n s_n)_{n=1}^{\infty}$ too converges but to rs.

Remark 25: In other words,

$$\lim_{n\to\infty}(r_n s_n) = \left(\lim_{n\to\infty} r_n\right)\left(\lim_{n\to\infty} s_n\right), \qquad (18)$$

provided both $\lim_{n\to\infty} r_n$ and $\lim_{n\to\infty} s_n$ exist. ■

Proof of Theorem 24

To prove this theorem means we must show that

$$\forall \varepsilon > 0, \quad \exists N \in \mathbb{N}, \quad \forall n \in \mathbb{N}, \; n > N \rightarrow | r_n s_n - rs | < \varepsilon . \tag{19}$$

Therefore, before going any further, it is well to first study the expression $| r_n s_n - rs |$ in order to find out more about it.

By the triangle inequality, we have

$$\begin{aligned} | r_n s_n - rs | &= | (r_n s_n - r_n s) + (r_n s - rs) | \\ &\leq | r_n s_n - r_n s | + | r_n s - rs | . \end{aligned} \tag{20}$$

Furthermore, we know that $(r_n)_{n=1}^{\infty}$ is convergent. Hence, by Theorem 20, it is bounded. Therefore, by Exercise 19, there is a fixed positive real number M such that

$$| r_n | \leq M \quad \text{for all } n \in \mathbb{N} . \tag{21}$$

Thus, from (20) and (21), we have

$$\begin{aligned} | r_n s_n - rs | &\leq | r_n s_n - r_n s | + | r_n s - rs | \\ &= | r_n (s_n - s) | + | s(r_n - r) | \\ &= | r_n | \, | s_n - s | + | s | \, | r_n - r | \\ &\leq M | s_n - s | + | s | \, | r_n - r |, \quad \text{for all } n \in \mathbb{N} . \end{aligned}$$

That is,

$$| r_n s_n - rs | \leq M | s_n - s | + | s | \, | r_n - r |, \quad \text{for all } n \in \mathbb{N} . \tag{22}$$

We are now in the position to prove (19). To do so, we will use the **method of exhaustion:**

Case 1. $s = 0$:

In this case (22) becomes

$$| r_n s_n - rs | \leq M | s_n |, \quad \text{for all } n \in \mathbb{N} . \tag{23}$$

Now, pick any real number $\varepsilon > 0$. Since $(s_n)_{n=1}^{\infty}$ converges to $s = 0$,

$$\exists N \in \mathbb{N}, \quad \forall n \in \mathbb{N}, \ n > N \to |\, s_n \,| < \tfrac{\varepsilon}{M}. \qquad (24)$$

Therefore, multiplying (24) by M and using inequality (23), we have:

$$\exists N \in \mathbb{N}, \quad \forall n \in \mathbb{N}, \ n > N \to |\, r_n s_n - rs \,| \le M |\, s_n \,| < \varepsilon,$$

which implies (19).

Case 2. $s \ne 0$:
Again let's pick any $\varepsilon > 0$. Then exploiting the fact that both sequences converge, we have:

$$\exists N_1 \in \mathbb{N}, \quad \forall n \in \mathbb{N}, \ n > N_1 \to |\, s_n - s \,| < \tfrac{\varepsilon}{2M} \qquad (25)$$

and

$$\exists N_2 \in \mathbb{N}, \quad \forall n \in \mathbb{N}, \ n > N_2 \to |\, r_n - r \,| < \tfrac{\varepsilon}{2|s|} \qquad (26)$$

Therefore, denoting by N the larger of the two natural numbers N_1 and N_2 we immediately obtain from (25) and (26) that

$$\forall n \in \mathbb{N}, \ n > N \to |\, r_n s_n - rs \,| < M \,|\, s_n - s \,| + |\, s \,|\,|\, r_n - r \,|$$

$$< \frac{\varepsilon}{2} + \frac{\varepsilon}{2} = \varepsilon,$$

which establishes (19). And the entire proof is thus complete. ♣

Example 26: Show that the limit of the following sequence is 0.

$$\left(\frac{2^n + 1}{2^n n} \right)_{n=1}^{\infty}.$$

Solution
First of all, note that

$$\frac{2^n + 1}{2^n n} = \frac{1}{n} \cdot \left[1 + \left(\tfrac{1}{2}\right)^n \right]$$

Hence, applying Theorem 24, we have

$$\lim_{n\to\infty}\frac{2^n+1}{2^n n} = \left(\lim_{n\to\infty}\frac{1}{n}\right)\cdot\left(\lim_{n\to\infty}\left[1+\left(\tfrac{1}{2}\right)^n\right]\right).$$

That is, by virtue of Theorem 21,

$$\lim_{n\to\infty}\frac{2^n+1}{2^n n} = 0\cdot\left(\lim_{n\to\infty}1+\lim_{n\to\infty}\left(\tfrac{1}{2}\right)^n\right).$$

Hence, in view of Example 13.1 and 13.3, we arrive at

$$\lim_{n\to\infty}\frac{2^n+1}{2^n n} = 0[1+0] = 0.$$ ♥

Corollary 27 [The Constant Factor Rule]: If c is any fixed real number and $(s_n)_{n=1}^{\infty}$ a sequence that converges to s, then the sequence $(c\cdot s_n)_{n=1}^{\infty}$ too converges but to cs.

Remark 28: In other words,

$$\lim_{n\to\infty}(c\cdot s_n) = c\left(\lim_{n\to\infty}s_n\right), \qquad (27)$$

provided $\lim_{n\to\infty}s_n$ exists ■

Proof of Corollary 27

All we need here is to notice that the sequence $(c\cdot s_n)_{n=1}^{\infty}$ is the product of the sequence $(s_n)_{n=1}^{\infty}$ and the constant sequence $(c)_{n=1}^{\infty}$. Indeed, having realized that, we may now apply Theorem 24 :

$$\lim_{n\to\infty}(c\cdot s_n) = \left(\lim_{n\to\infty}c\right)\left(\lim_{n\to\infty}s_n\right),$$

which, in light of Example 13.1, reduces to

$$\lim_{n\to\infty}(c\cdot s_n) = c\left(\lim_{n\to\infty}s_n\right),$$

as we sought to prove . ♣

The following is a lemma that, though not in itself interesting, will be very useful in establishing our next theorem.

Lemma 29: Let $(s_n)_{n=1}^{\infty}$ converge to a nonzero real number s. Then, there is a number $N \in \mathbb{N}$ such that for all $n > N$,

$$\frac{1}{|s_n|} \leq \frac{2}{|s|} .$$

Proof

Since $(s_n)_{n=1}^{\infty}$ converges to the nonzero real number s,

$$\exists N \in \mathbb{N}, \quad \forall n \in \mathbb{N}, \; n > N \rightarrow |s_n - s| < \tfrac{|s|}{2}. \qquad (28)$$

Therefore, applying Corollary 10.3 of Chapter 14 to (28), we have

$$\exists N \in \mathbb{N}, \quad \forall n \in \mathbb{N}, \; n > N \rightarrow ||s_n| - |s|| < \tfrac{|s|}{2}.$$

That is

$$\exists N \in \mathbb{N}, \quad \forall n \in \mathbb{N}, \; n > N \rightarrow |s| - \tfrac{|s|}{2} < |s_n| < |s| + \tfrac{|s|}{2}.$$

Hence,

$$\exists N \in \mathbb{N}, \quad \forall n \in \mathbb{N}, \; n > N \rightarrow \tfrac{|s|}{2} < |s_n|,$$

which yields

$$\frac{1}{|s_n|} \leq \frac{2}{|s|} \quad \text{for all } n > N,$$

the result we wished to establish. ♣

Theorem 30 [The quotient Rule]: If $(r_n)_{n=1}^{\infty}$ converges to r and $(s_n)_{n=1}^{\infty}$ converges to a **nonzero** real s, then the **quotient sequence** $(r_n / s_n)_{n=1}^{\infty}$ too converges to r/s.

Remark 31: In other words,

$$\lim_{n\to\infty} \frac{r_n}{s_n} = \frac{\lim\limits_{n\to\infty} r_n}{\lim\limits_{n\to\infty} s_n},$$

provided both $\lim\limits_{n\to\infty} r_n$ and $\lim\limits_{n\to\infty} s_n$ exists and $\lim\limits_{n\to\infty} s_n$ is nonzero. ■

Proof of Theorem 30

To prove this theorem means we must show that

$$\forall \varepsilon > 0, \quad \exists N \in \mathbb{N}, \quad \forall n \in \mathbb{N}, \ n > N \to |\, r_n / s_n - r / s \,| < \varepsilon . \qquad (29)$$

Therefore, before going any further, it is worth studying the expression

$$|\, r_n / s_n - r / s \,| .$$

We have

$$|\, r_n / s_n - r / s \,| = \left| \frac{r_n s - s_n r}{s_n s} \right| = \left| \frac{s(r_n - r) + (s - s_n) r}{s_n s} \right| . \qquad (30)$$

Hence, using the triangle inequality,

$$|\, r_n / s_n - r / s \,| \leq \frac{|\, s(r_n - r)\,| + |\,(s - s_n) r\,|}{|\, s_n s \,|}$$

$$= \frac{|\, r_n - r \,|}{|\, s_n \,|} + \frac{|\, r \,|}{|\, s \,|} \cdot \frac{|\, s - s_n \,|}{|\, s_n \,|} .$$

Therefore, using Lemma 29, we see that there is a natural number N such that for all $n > N$,

$$|\, r_n / s_n - r / s \,| \leq \frac{2}{|\, s \,|} \cdot |\, r - r_n \,| + \frac{2 |\, r \,|}{|\, s \,|^2} \cdot |\, s - s_n \,| . \qquad (31)$$

Now, we are in the position to prove (29) using (31). To do so, we shall distinguish two cases.

Case 1. $r = 0$:

In this case (31) becomes

$$|\, r_n / s_n - r / s \,| \leq \frac{2}{|\, s \,|} \cdot |\, r_n \,|, \quad \text{for all } n > N . \qquad (32)$$

So, for every $\varepsilon > 0$, since $(r_n)_{n=1}^{\infty}$ converges to $r = 0$,

$$\exists N_1 \in \mathbb{N}, \quad \forall n \in \mathbb{N},\ n > N_1 \rightarrow | r_n | < \tfrac{|s|\varepsilon}{2}, \tag{33}$$

Therefore, denoting by N_2 the larger of the two natural numbers N_1 and N, we easily deduce from (32) and (33) that

$$\forall n \in \mathbb{N},\ n > N_2 \rightarrow \ | r_n / s_n - r / s | \leq \frac{2}{|s|} \cdot | r_n | < \varepsilon,$$

which gives us the result we sought out to establish.

Case 2. $r \neq 0$:
For every $\varepsilon > 0$, we have

$$\exists N_1 \in \mathbb{N}, \quad \forall n \in \mathbb{N},\ n > N_1 \rightarrow | s_n - s | < \frac{|s|^2}{4|r|}\varepsilon \tag{34}$$

and

$$\exists N_2 \in \mathbb{N}, \quad \forall n \in \mathbb{N},\ n > N_2 \rightarrow | r_n - r | < \tfrac{|s|\varepsilon}{4} \tag{35}$$

Therefore, denoting by N_3 the largest of the three natural numbers N_1, N_2 and N, we conclude from (31), (34) and (35) that,

$$n > N_3 \rightarrow | r_n / s_n - r / s | \leq \frac{2}{|s|} \cdot | r - r_n | + \frac{2|r|}{|s|^2} \cdot | s - s_n | < \tfrac{\varepsilon}{2} + \tfrac{\varepsilon}{2} = \varepsilon$$

which again gives us the desired result. Hence, the whole proof is complete. ♣

Example 31: Show that the limit of the following sequence is $2/3$.

$$\left(\frac{2n^2 + 3n}{3n^2 + 2n} \right)_{n=1}^{\infty}.$$

Solution
First of all, we re-write the general term s_n of our sequence by dividing both its numerator and denominator by the leading monomial n^2, to obtain

$$\frac{2n^2 + 3n}{3n^2 + 2n} = \frac{2 + 3/n}{3 + 2/n}.$$

Hence, applying Theorem 30, we have:

$$\lim_{n\to\infty} \frac{2n^2 + 3n}{3n^2 + 2n} = \lim_{n\to\infty} \frac{2 + 3/n}{3 + 2/n} = \frac{\lim_{n\to\infty}(2 + 3/n)}{\lim_{n\to\infty}(3 + 2/n)} = \frac{2 + 0}{3 + 0} = 2/3.$$ ♥

PROBLEMS

1. Show that if $(s_n)_{n=1}^{\infty}$ converges to l, then the sequence $(s_{n-1})_{n=2}^{\infty}$ also converges to l.

2. Let p be any fixed natural number. Show that if the sequence $(s_n)_{n=1}^{\infty}$ converges to l, then the sequence $(s_{n-p})_{n=p+1}^{\infty}$ also converges to l.

3. Show that if a sequence of positive terms $(s_n)_{n=1}^{\infty}$ converges to l, then $\left(\sqrt{s_n}\right)_{n=1}^{\infty}$ too converges $\sqrt{l}$.

4.* Show that if $(n_k)_{k=1}^{\infty}$ is a sequence of natural numbers satisfying $n_k \le n_{k+1}$ for all $k \in \mathbb{N}$, then $n_k \ge k$ for all $k \in \mathbb{N}$.

5.* Let $f : \mathbb{R} \to \mathbb{R}$ be a function satisfying Condition 6.1 of Problem 6 in Chapter 12. Then show that if a sequence $(s_n)_{n=1}^{\infty}$ converges to l, then the corresponding sequence $\left(f(s_n)\right)_{n=1}^{\infty}$ also converges but to $f(l)$.

6. Prove that $(s_n)_{n=1}^{\infty}$ converges to s if and only if $(s_n - s)_{n=1}^{\infty}$ converges to 0.

7. Show that if $(s_n)_{n=1}^{\infty}$ converges to s, then $(|s_n|)_{n=1}^{\infty}$ converges $|s|$.

8. **[The Squeeze Theorem].** Assume that $(s_n)_{n=1}^{\infty}$, $(x_n)_{n=1}^{\infty}$ and $(r_n)_{n=1}^{\infty}$ are such that both $(s_n)_{n=1}^{\infty}$ and $(r_n)_{n=1}^{\infty}$ converge to a same limit l. Furthermore, assume that the inequalities $s_n \le x_n \le r_n$ hold true for any $n \in \mathbb{N}$. Then prove that $(x_n)_{n=1}^{\infty}$ too converges to l.

"There are no deep theorems - only theorems that we have not understood very well."
The Mathematical Intelligencer, vol. 5, no. 3, 1983.

CHAPTER 17

FUNDAMENTAL[21] SEQUENCES, AND MONOTONIC SEQUENCES

17.1. FUNDAMENTAL SEQUENCES

DEFINITIONS

In Chapter 16, after introducing the idea of a convergent sequence, we presented a number of worked examples in each of which we showed that the sequence converges to a known limit. To do so in each case, we first of all had to know [or at least suspect] the value of the limit l , and then show that Definition 8 or 10 was satisfied by the sequence and its limit. Unfortunately, in most cases the limits l are unknown in advance, and so it is impossible to use in these cases either Definition 8 or Definition 10. Hence, what is most needed is a criterion for convergence that does not require the explicit knowledge of the limit l .

[21] **Fundamental** sequences are also called **Cauchy** sequences.

Before giving such a criterion [that applies to all sequences], we will first introduce the following

Definition 1 [Fundamental sequences]: A sequence $(s_n)_{n=1}^{\infty}$ is said to be **fundamental**, if

$$\begin{gathered}\forall \varepsilon > 0, \quad \exists N \in \mathbb{N}, \\ \forall n \in \mathbb{N}, \quad \forall m \in \mathbb{N}, \quad (n > N \text{ and } m > N) \rightarrow |s_m - s_n| < \varepsilon .\end{gathered} \tag{1}$$

Since for any two natural numbers n and m, one of them, say m, is always bigger than or equal to the other n, we can write $m = n + p$, where p is some whole number. Thus the foregoing definition can be rewritten as

Definition 2 [Fundamental sequences]: A sequence $(s_n)_{n=1}^{\infty}$ is said to be fundamental, if

$$\forall \varepsilon > 0, \quad \exists N \in \mathbb{N}, \quad \forall n \in \mathbb{N}, \quad \forall p \in \mathbb{W}, \quad n > N \rightarrow |s_{n+p} - s_n| < \varepsilon ,$$

where $\mathbb{W} = \{0\} \cup \mathbb{N}$.

With the foregoing definitions, we are now in the position to introduce two theorems, the latter being the criterion for convergence we are looking for.

PROPERTIES

Theorem 3: if $(s_n)_{n=1}^{\infty}$ is fundamental, then it is bounded.

Proof

Assume $(s_n)_{n=1}^{\infty}$ is a fundamental sequence. Then by choosing $\varepsilon = 1$ in Definition 1, we have

$$\exists N \in \mathbb{N}, \quad \forall n \in \mathbb{N}, \quad \forall m \in \mathbb{N}, \quad (n > N \text{ and } m > N) \rightarrow |\, s_n - s_m \,| < 1$$

Thus allowing $m = N + 1$, a natural number clearly greater than N, we get

$$\exists N \in \mathbb{N}, \quad \forall n \in \mathbb{N}, \quad n > N \rightarrow |\, s_n - s_{N+1} \,| < 1, \qquad (2)$$

which, can also be written as

$$\exists N \in \mathbb{N}, \quad \forall n \in \mathbb{N}, \quad n > N \rightarrow -1 < s_n - s_{N+1} < 1. \qquad (3)$$

Therefore

$$\exists N \in \mathbb{N}, \quad \forall n \in \mathbb{N}, \quad n > N \rightarrow -1 + s_{N+1} < s_n < s_{N+1} + 1. \qquad (4)$$

Thus, the sequence $(s_n)_{n=1}^{\infty}$ is bounded from above by the biggest of the real numbers $s_1, s_2, \ldots, s_N$ and $s_{N+1} + 1$, and from below by the smallest of the real numbers $s_1, s_2, \ldots, s_N$ and $-1 + s_{N+1}$. ♣

We will now establish a famous criterion for convergence. This criterion will not require knowledge of the limit of the sequence. The criterion is commonly known as Cauchy convergence criterion; it is named after the great French mathematician **Augustin Cauchy (1789-1857)**.

Theorem 4 [Cauchy convergence criterion]: A sequence $(s_n)_{n=1}^{\infty}$ is convergent, if and only if it is fundamental.

Proof

To prove Theorem 4 requires that we prove both of these implications:

1. $(s_n)_{n=1}^{\infty}$ converges $\Rightarrow$ $(s_n)_{n=1}^{\infty}$ is fundamental,

and

2. $(s_n)_{n=1}^{\infty}$ is fundamental $\Rightarrow$ $(s_n)_{n=1}^{\infty}$ is converges.

Proof of implication 1. [This is the easier part of the proof]

Suppose that $(s_n)_{n=1}^{\infty}$ converges to the limit l. Then for every real number $\varepsilon > 0$, there exists a natural number N such that

$$\forall n \in \mathbb{N}, \quad n > N \to |s_n - l| < \tfrac{\varepsilon}{2} \tag{5}$$

and

$$\forall m \in \mathbb{N}, \quad m > N \to |s_m - l| < \tfrac{\varepsilon}{2}. \tag{6}$$

Hence, using the triangle inequality seen in Chapter 14, we deduce from (5) and (6) that:

$$\begin{aligned}(n > N \text{ and } m > N) \to |s_m - s_n| &= |(s_m - l) + (l - s_n)| \\ &\leq |s_m - l| + |l - s_n| \\ &< \frac{\varepsilon}{2} + \frac{\varepsilon}{2} = \varepsilon .\end{aligned}$$

Thus,

$$\forall \varepsilon > 0, \quad \exists N \in \mathbb{N},$$
$$\forall n \in \mathbb{N}, \quad \forall m \in \mathbb{N}, \quad (n > N \text{ and } m > N) \to |s_m - s_n| < \varepsilon ,$$

which is the same thing as saying that $(s_n)_{n=1}^{\infty}$ is fundamental. [See Definition 1.]

Proof of implication 2. [This is the harder part of the proof.]

Assume that $(s_n)_{n=1}^{\infty}$ is fundamental. Now, let us denote by S the set of real numbers less than all but (perhaps) a finite number of terms of the sequence $(s_n)_{n=1}^{\infty}$:

$$S = \{x \in \mathbb{R} : \exists n_x \in \mathbb{N}, \ \forall n > n_x, \ x < s_n\}.$$

It is not difficult to show that the set S is nonempty. Indeed, since the sequence $(s_n)_{n=1}^{\infty}$ is fundamental, it is bounded. That is, there are two fixed real numbers A and B such that

$$\forall n \in \mathbb{R}, \quad A < s_n < B, \tag{7}$$

which shows that $A \in S$. Thus S is nonempty.

Also, since every element of S is smaller than some term of the sequence $(s_n)_{n=1}^{\infty}$, the second inequality in (7) implies that the

set S is bounded from above by B. [Is set S bounded from below?] Hence, by the least upper bound principle, the nonempty set S, bounded from above, admits a least upper bound, say s.
We will now show that this s is the limit of our fundamental sequence $(s_n)_{n=1}^{\infty}$. For that, let us pick any $\varepsilon > 0$. Since $(s_n)_{n=1}^{\infty}$ is fundamental,

$$\exists N \in \mathbb{N}, \ \forall n \in \mathbb{N}, \ \ \forall m \in \mathbb{N}, \\ (n > N \text{ and } m > N) \to | s_m - s_n | < \varepsilon . \tag{8}$$

In particular, for the same natural number N chosen in (8), we have

$$\forall n \in \mathbb{N}, \quad n > N \to | s_n - s_{N+1} | < \tfrac{\varepsilon}{2}, \tag{9}$$

which may be re-written as

$$\forall n \in \mathbb{N}, \quad n > N \to -\tfrac{\varepsilon}{2} < s_n - s_{N+1} < \tfrac{\varepsilon}{2} . \tag{10}$$

From (10), we also have

$$\forall n \in \mathbb{N}, \quad n > N \to -\tfrac{\varepsilon}{2} + s_{N+1} < s_n .$$

Therefore, the number $-\frac{\varepsilon}{2} + s_{N+1}$ belongs to the set S. Thus,

$$-\tfrac{\varepsilon}{2} + s_{N+1} \leq s . \tag{11}$$

Again, from (10), we also have

$$\forall n \in \mathbb{N}, \quad n > N \to s_n < \tfrac{\varepsilon}{2} + s_{N+1} . \tag{12}$$

Thus, $\frac{\varepsilon}{2} + s_{N+1}$ does not belong to the set S. [For $\frac{\varepsilon}{2} + s_{N+1}$ is not less than an infinite number of terms of the sequence.]

Therefore ,

$$s \leq \tfrac{\varepsilon}{2} + s_{N+1} . \tag{13}$$

[For, if $\frac{\varepsilon}{2} + s_{N+1}$ were less than s, then it would also be less than an infinite number of elements of $(s_n)_{n=1}^{\infty}$, which would contradict (12).]

Now, from (11) and (13), we obtain

$$|s - s_{N+1}| < \frac{\varepsilon}{2}. \tag{14}$$

Finally, by invoking (9), (14) together with the triangle inequality, we have

$$\forall n \in \mathbb{N},\ n > N \rightarrow |s_n - s| < |s_n - s_{N+1}| + |(s_{N+1} - s|$$
$$< \frac{\varepsilon}{2} + \frac{\varepsilon}{2} = \varepsilon,$$

which establishes the convergence of $(s_n)_{n=1}^{\infty}$ to s. And so, the entire proof is complete. ♣

Example 5: Assume we are given:

1. a function $f : \mathbb{R} \rightarrow \mathbb{R}$ such that for every x and y in $\mathbb{R}$,

$$|f(x) - f(y)| \leq r|x - y|, \qquad 0 < r < 1.$$

2. and a sequence $(s_n)_{n=1}^{\infty}$ defined recursively by

$$s_1 \in \mathbb{R},\ \textit{and}\ \ s_n = f(s_{n-1})\ \ \textit{for all}\ \ n \geq 2.$$

Then show that $(s_n)_{n=1}^{\infty}$ is a convergent sequence. [Hint: Use Problem 6 of Chapter 12.]

Solution

Note that, in this example, we have no knowledge of the limit of our sequence $(s_n)_{n=1}^{\infty}$. It is therefore impossible to prove the convergence of $(s_n)_{n=1}^{\infty}$ directly by checking to see if either Definition 8 or 10 of the last chapter is satisfied. Hence, what we will do here is to show that our sequence is fundamental; and by the foregoing Cauchy theorem, this will suffice.

Let $n \in \mathbb{N}$ and $p \in \mathbb{W}$. We have:

$$|s_{n+p} - s_n|$$

$$= |(s_{n+p} - s_{n+[p-1]}) + (s_{n+[p-1]} - s_{n+[p-2]}) + \ldots + (s_{n+(p-(p-1))} - s_n)|$$

$$\le |s_{n+p} - s_{n+p-1}| + |s_{n+p-1} - s_{n+p-2}| + \ldots + |s_{n+p-(p-1)} - s_n|$$

$$= |s_{n+p} - s_{n+p-1}| + |s_{n+p-1} - s_{n+p-2}| + \ldots + |s_{n+1} - s_n|$$

$$\le r^{n+p-2} |s_2 - s_1| + r^{n+p-3} |s_2 - s_1| + \ldots + r^{n-1} |s_2 - s_1|$$

$$= \left(r^{p-1} + r^{p-2} + \ldots + r^0\right) r^{n-1} |s_2 - s_1|.$$

$$= \frac{1 - r^p}{1 - r} r^{n-1} |s_2 - s_1|.$$

$$\le C r^n, \quad \text{where } C = |s_2 - s_1| / [r(1-r)].$$

Hence, using the result of Example 12.3 of Chapter 16, we may conclude that

$$\forall \varepsilon > 0, \quad \exists N \in \mathbb{N},$$

$$\forall n \in \mathbb{N}, \ \forall p \in \mathbb{W}, \quad n > N \to |s_{n+p} - s_n| < C r^n < \varepsilon.$$

Therefore, $(s_n)_{n=1}^{\infty}$ is a fundamental sequence; and so, by Theorrem 4, it is a convergent sequence. ♥

17.2. MONOTONIC SEQUENCES

We are now going to look at another kind of sequences called **monotonic** sequences. The reason we are looking at these sequences is that, in addition to their importance in mathematics and everyday life, there is also for these sequences a simple criterion for convergence that does not require the explicit knowledge of their limits. To define the idea of monotonic sequences, we will first need to introduce the concepts of increasing, decreasing, nondecreasing, and nonincreasing sequences.

Definition 6 [Increasing Sequences]: A sequence $(s_n)_{n=1}^{\infty}$ is said to be **increasing**, if the inequality

$$s_n < s_{n+1}$$

holds for all $n \in \mathbb{N}$.

Example 7: Show that the sequence $(a_n)_{n=1}^{\infty}$ defined by

$$a_n = \sqrt{n} - \sqrt{n+1}.$$

is increasing.

Solution

To show that $(a_n)_{n=1}^{\infty}$ is an increasing sequence, we first evaluate the sign of the difference:

$$\begin{aligned} a_{n+1} - a_n &= \left[\sqrt{n+1} - \sqrt{n+2}\right] - \left[\sqrt{n} - \sqrt{n+1}\right] \\ &= \frac{-1}{\sqrt{n+1} + \sqrt{n+2}} - \frac{-1}{\sqrt{n} + \sqrt{n+1}} \\ &= \frac{\sqrt{n+2} - \sqrt{n}}{\left[\sqrt{n} + \sqrt{n+1}\right]\left[\sqrt{n+1} + \sqrt{n+2}\right]} > 0 \end{aligned}$$

Thus,

$$a_{n+1} > a_n, \quad \text{for all } n \in \mathbb{N}.$$

That is, the sequence $(a_n)_{n=1}^{\infty}$ is indeed increasing. ♥

Definition 8 [Decreasing Sequences]: A sequence $(s_n)_{n=1}^{\infty}$ is said to be **decreasing**, if the inequality

$$s_{n+1} < s_n$$

holds for all $n \in \mathbb{N}$.

Example 9: Show that the sequence $(u_n)_{n=1}^{\infty}$ defined by

$$u_n = \sqrt{n+1} - \sqrt{n}.$$

is decreasing.

Solution

Since this is trivial, we leave it to the reader to supply all the details. ♥

Definition 10 [Nondecreasing Sequences]: A sequence $(s_n)_{n=1}^{\infty}$ is said to be **nondecreasing**, if the inequality

$$s_n \leq s_{n+1}$$

holds for all $n \in \mathbb{N}$.

Example 11: Show that the sequence

$(a_n)_{n=1}^{\infty} = (5, 6, 6, 7, 7, 7, 8, 8, 8, 8, \ldots)$ is nondecreasing.

Solution

Since it is trivial, we leave it to the reader to supply all the details. ♥

Definition 12 [Nonincreasing Sequences]: A sequence $(s_n)_{n=1}^{\infty}$ is said to be **nonincreasing**, if the inequality

$$s_{n+1} \leq s_n$$

holds for all $n \in \mathbb{N}$.

Example 13: Show that the sequence

$(a_n)_{n=1}^{\infty} = (5, 5, 4, 4, 4, 3, 3, 3, 3, \ldots)$ is nonincreasing.

Solution

Since this one too is trivial, we leave it to the reader to supply all the details. ♥

We are now in the position to introduce the concept of a monotonic sequence.

Definition 14 [Monotonic Sequences]: A sequence $(s_n)_{n=1}^{\infty}$ is said to be **monotonic**, if it is increasing, decreasing, nondecreasing, or nonincreasing.

The following is a criterion for convergence of monotonic sequences. Like the Cauchy convergence criteria, this criterion too does not require an explicit knowledge of the limit. It is named after the great German mathematician **Weierstrass (1815-1897)**.

Theorem 15 [Weierstrass' convergence criterion]: Assume sequence $(s_n)_{n=1}^{\infty}$ is a monotonic sequence. Then $(s_n)_{n=1}^{\infty}$ is converges, if and only if it is bounded.

Proof

To prove this theorem requires that we prove both of these implications:

1. $(s_n)_{n=1}^{\infty}$ converges $\Rightarrow$ $(s_n)_{n=1}^{\infty}$ is bounded,

and

2. $(s_n)_{n=1}^{\infty}$ is bounded $\Rightarrow$ $(s_n)_{n=1}^{\infty}$ converges.

Proof of implication 1.

Suppose that $(s_n)_{n=1}^{\infty}$ converges. Then, from Theorem 20 of Chapter16, $(s_n)_{n=1}^{\infty}$ is also bounded. And so the proof of this part is complete.

Proof of implication 2.

Now, assume the sequence $(s_n)_{n=1}^{\infty}$ is monotonic and bounded.

We may also assume that $(s_n)_{n=1}^{\infty}$ is increasing since, as will be evident, the arguments developed in this case apply as well to the three other cases of monotonicity.

Now, from the boundedness of $(s_n)_{n=1}^{\infty}$ we know that there are fixed real numbers A and B such that

$$\forall n \in \mathbb{R}, \quad A < s_n < B. \tag{15}$$

Therefore, the set S defined by

$$S = \{ \, s_n : \; n \in \mathbb{N} \, \}$$

is bounded. Hence, by Theorem 19 of Chapter 11, S has a least upper bound, say s. Now, let ε be any positive real number. Since $s - \varepsilon$ is smaller s, we know that $s - \varepsilon$ is not an upper bound of S. [For s is the least upper bound.] Therefore, there is at least one element of S, say s_N, such that

$$s - \varepsilon < s_N . \tag{16}$$

Hence, using our assumption that $(s_n)_{n=1}^{\infty}$ is an increasing sequence, we have

$$\forall n \in \mathbb{N}, \quad n > N \to s - \varepsilon < s_N < s_n \le s < s + \varepsilon .$$

Thus, we have proved that

$$\forall \varepsilon > 0, \quad \exists N \in \mathbb{N},$$
$$\forall n \in \mathbb{N}, \; n > N \to s - \varepsilon < s_n < s + \varepsilon .$$

which completes the proof of this second part.

And so is the proof of the entire theorem is also complete. ♣

Example 16: Prove that the sequence $(a_n)_{n=1}^{\infty}$ defined recursively by

$$a_1 = \sqrt{2}, \qquad a_n = \sqrt{2 + \sqrt{a_{n-1}}} \quad \text{for } n \ge 2$$

converges; and show that its limit satisfies the equation

$$l^4 - 4l^2 - l + 4 = 0 .$$

Solution

First, let us find the first three terms of the sequence:

$$a_1 = \sqrt{2}, \quad a_2 = \sqrt{2 + \sqrt{2}}, \quad a_3 = \sqrt{2 + \sqrt{2 + \sqrt{2}}} .$$

We see that $a_1 < a_2 < a_3$. Thus it looks as though the sequence $(a_n)_{n=1}^{\infty}$ is increasing and bounded from above by 2. We are going to show that this is indeed the case by mathematical induction.

So, let us denote by $P(n)$ the open sentence:

$$a_n < a_{n+1} \quad \text{and} \quad a_n < 2\,.$$

Initial Step. From the above we see that the proposition $P(1)$ is:

$$\sqrt{2} < \sqrt{2+\sqrt{2}} \quad \text{and} \quad \sqrt{2} < 2.$$

Thus, $P(1)$ is true.

Induction hypothesis. Assume $P(k)$ is true for some $k \in \mathbb{N}$. That is, assume

$$a_k < a_{k+1} \quad \text{and} \quad a_k < 2\,,$$

is true for some $k \in \mathbb{N}$.

Inductive step. Now, we will deduce from the *Inductive hypothesis* that $P(k+1)$ too is true. Using the induction hypothesis, we have:

$$a_{k+2} = \sqrt{2+\sqrt{a_{k+1}}} > \sqrt{2+\sqrt{a_k}} = a_{k+1}$$

and

$$a_{k+1} = \sqrt{2+\sqrt{a_k}} < \sqrt{2+2} = 2\,,$$

Thus, assuming that $P(k)$ is true makes $P(k+1)$ too true.

Therefore, by Theorem 2 of Chapter 12, we have shown that $a_n < a_{n+1}$ and $a_n < 2$, for all $n \in \mathbb{N}$. Hence the sequence is increasing and bounded from above. Thus, by Weierstrass convergence criterion, it has a limit, say l.

It now remains to show that l verifies the equation $l^4 - 4l^2 - l + 4 = 0$.

From the fact that

$$(a_n)_{n=1}^{\infty} = \left(\sqrt{2 + \sqrt{a_{n-1}}}\right)_{n=1}^{\infty},$$

we have

$$\lim a_n = \lim\left(\sqrt{2 + \sqrt{a_{n-1}}}\right),$$

which, by a repeated use of the result of Problems 3 and 1 of Chapter 16, gives

$$\lim a_n = \sqrt{\lim 2 + \sqrt{\lim a_{n-1}}}.$$

That is

$$l = \sqrt{2 + \sqrt{l}}.$$

Hence, by squaring, we get

$$l^4 - 4l^2 - l + 4 = 0. \heartsuit$$

PROBLEMS

1. Prove that $(s_n)_{n=1}^{\infty}$ converges to s, if and only if $(s_n - s)_{n=1}^{\infty}$ converges to 0.

2. Prove that if $(s_n)_{n=1}^{\infty}$ and $(a_n)_{n=1}^{\infty}$ are fundamental sequences, then $(s_n + a_n)_{n=1}^{\infty}$ too is a fundamental sequence.

3. Show that the limit s of the sequence $(s_n)_{n=1}^{\infty}$ in Example 5 satisfies the condition: $s = f(s)$.

4. Prove that if $(s_n)_{n=1}^{\infty}$ and $(a_n)_{n=1}^{\infty}$ are fundamental sequences, then $(s_n a_n)_{n=1}^{\infty}$ too is a fundamental sequence.

5. Prove directly that $\left(\frac{2n+5}{n}\right)_{n=1}^{\infty}$ is a fundamental sequence.

6. Let a_1 and a_2 be distinct real numbers. Now define $a_n = (a_{n-1} + a_{n-2})/2$ for each natural number $n \geq 3$. Prove that $(a_n)_{n=1}^{\infty}$ is a fundamental sequence. Hint: You may first establish by induction that $a_n - a_{n-1} = (a_2 - a_1)/(-2)^{n-2}$.

7. Let $a_1 = 2$, and $a_n = (a_{n-1} + 1)/2$ for each natural number $n \geq 2$. Prove that $(a_n)_{n=1}^{\infty}$ is a decreasing sequence, then find its limit. [Hint: First, show that $a_n > 1$ *for all* $n \in \mathbb{N}$.]

8. Let $a_1 = \sqrt{2}$. Now define $a_n = \sqrt{2 + a_{n-1}}$ for each natural number $n \geq 2$. Prove that the sequence $(a_n)_{n=1}^{\infty}$ converges, and find its limits.

"Our minds are finite, and yet even in these circumstances of finitude we are surrounded by possibilities that are infinite, and the purpose of life is to grasp as much as we can out of that infinitude."
Alfred North Whitehead.

CHAPTER 18

FINITE AND INFINITE SETS

In this Chapter we will discuss the ideas of finite and infinite sets. We will also study some of the basic properties of these sets. We will see that, unlike finite sets, infinite sets enjoy properties that seem counterintuitive. As the above quote suggests, infinite sets are very important in both mathematics and everyday life.

18.1. FINITE SETS

Most people already have an idea of what a finite set is. Indeed, when asked to definite a finite set, it is likely they will say that a finite set is one whose elements can all be counted with the counting process eventually coming to an end [at some natural number K.] Mathematically, this may be translated as follows: Given any nonempty set S, this set is finite, if there is some natural number K such that the elements of S and those of set $\{1, 2, 3, \ldots, K\}$ can be paired off in a one-to-one correspondence. Hence

Definition1 [Finite Sets]: A set S is said to be **finite**, if

1. either $S = \varnothing$,
2. or there is a number $K \in \mathbb{N}$ such that a bijection

$$f : \{1, 2, 3, \ldots, K\} \to S$$

exists.

In the case of $S = \varnothing$, we say that the **size** or **cardinality** or **number of elements** of set S is zero; and we write $\operatorname{card}(S) = 0$. On the other hand, when S is nonempty, the cardinality of S is taken to be the natural number K, and we write $\operatorname{card}(S) = K$.

Remark 2:
From the foregoing definition, it is easy to see that a nonempty set S is finite, if and only if the totality of its elements can be listed as: $f(1)$, $f(2)$, $f(3)$, ... $f(K)$, for some **fixed** natural number $K \in \mathbb{N}$. ■

Hence, by letting

$$\begin{aligned} f_1 &= f(1), \\ f_2 &= f(2), \\ f_3 &= f(3), \\ &\cdots\cdots\cdots \\ &\cdots\cdots\cdots \\ &\cdots\cdots\cdots \\ f_K &= f(K), \end{aligned}$$

we may replace Definition 1 by yet another:

Definition 3 [Finite Sets]: A set S is **finite**, if
1. either $S = \varnothing$,
2. or S can be listed as $\{f_1, f_2, f_3, \ldots, f_K\}$ for some fixed $K \in \mathbb{N}$.

Example 4: Prove the following propositions.

1. The set $\{a, b, c, d\}$ is finite.
2. The set of natural numbers is not finite.

Solution
1. The set $\{a, b, c, d\}$ is finite. Indeed, by letting $f : \{1, 2, 3, 4\} \to \{a, b, c, d\}$ be the bijection defined by:

$$f(1) = a, \quad f(2) = b, \quad f(3) = c \quad \text{and} \quad f(4) = d,$$

it is obvious that $\{a, b, c, d\}$ can be written in the form: $\{f_1, f_2, f_3, f_4\}$.

2. The set of natural numbers $\mathbb{N}$ is not finite. We can prove this by **contradiction**. Indeed, let us assume that the proposition is false; that is to say, suppose $\mathbb{N}$ is finite. Then, by Definition 2, there is a natural number K such that $\mathbb{N}$ can be listed as

$$\mathbb{N} = \{f_1, f_2, f_3, \ldots, f_K\} . \tag{1}$$

Now, let us consider the sum of all the elements of $\mathbb{N}$:

$$f = f_1 + f_2 + f_3 + \ldots + f_K .$$

Since f is a sum of natural numbers, f itself must be a natural number; $f \in \mathbb{N}$. But, as a sum of nonzero numbers, we must have f greater than any of its terms. Hence:

$$f \neq f_n, \quad \forall n \in \{1, 2, 3, \ldots, K\} .$$

Therefore, $f \notin \mathbb{N}$, which clearly contradicts our earlier finding that $f \in \mathbb{N}$. ♥

The following theorem seems so obvious that for most people it hardly requires any proof.

Theorem 5: Assume A is a finite set. Then any subset S of A is also finite. Furthermore, if S is a proper subset of A, then $\text{card}(S) < \text{card}(A)$.

Proof

Two cases may be distinguished:

Case 1. The set S is a trivial subset of A:
In this case $S = \varnothing$ or $S = A$. Hence, by Definition 3, S is finite.

Case 2. The set S is a subset of A but not a trivial one:

First, you may recall that in this case the set S may be obtained from A by deleting some of the elements of A [but not all of them.] Now, since A is finite, we may list it as follows:

$$A = \{a_1, a_2, a_3, \ldots, a_K\} \tag{2}$$

Furthermore, we will assume that in (2) those items that do not belong to S all appear at the end of the listing. [For, indeed, such an arrangement of the element of a finite set A is always feasible.]

Therefore, by deleting from (2) all the items that do not belong to S we obtain:

$$S = \{a_1, a_2, a_3, \ldots, a_J\} \quad \textit{with} \quad J < K$$

Hence, S too is a finite set with cardinality less than that of its superset A. ♣

Next, we will establish a very important theorem, called the Pigeonhole principle. The reason this theorem is called **the Pigeonhole principle** is that one may interpret it as follows: *If one tries to pair off a fixed number of pigeons with a less number of pigeonholes, then sooner or later one will have to assign more than one pigeon to at least one hole.*

Theorem 6 [Pigeonhole principle]: Assume R and S are nonempty finite sets. If $\text{card}(R) > \text{card}(S)$, then there is no injection defined on R into S.

Proof

We will prove the present theorem by **mathematical induction** on the size of the codomain S.

So, let us denote by $P(n)$ the open sentence:

If $\text{card}(R) > \text{card}(S) = n$, then there is no injection on R into S.

Initial step. From the above, we see that $P(1)$ is the proposition:

If R contains more than one item, and S exactly one item, say $S = \{s\}$; then there is no injection from R into S.

So $P(1)$ is true. For, indeed, if $f : R \to S$ is a function from R into S then all elements of R, which number more than one, are mapped on the same item $s \in S$. Thus, there are a least two distinct elements of R, say r_1 and r_2 such that $f(r_1) = s = f(r_2)$. Hence, f is not injective.

Induction hypothesis. Assume $P(k)$ is true for some natural number k. That is, assume the proposition

If $\text{card}(R) > \text{card}(S) = k$, *then there is no injection from* R *into* S

is true for some natural number k.

Inductive step. Now, we will show that the truth of $P(k+1)$ follows from the induction hypothesis.
Let $f : R \to S$ and $\text{card}(R) > \text{card}(S) = k+1$. Since S has size $k+1$, we may list it as follows:

$$S = \{s_1, s_2, s_3, \ldots, s_{k+1}\}$$

Now, let us fix in S one item, say s_{i_0}. We will distinguish three cases:

Case 1. The element s_{i_0} has no pre-image in R. In this case, let us introduce a new function $g : R \to S \setminus \{s_{i_0}\}$ defined by $g(r) = f(r)$ for every $r \in R$. But then $\text{card}(R) > \text{card}\left(S \setminus \{s_{i_0}\}\right) = k$. Therefore, by our inductive hypothesis, g is not injective. Hence, $g(r_1) = g(r_2)$ for some distinct elements $r_1 \in R$ and $r_2 \in R$, which in turn means that $f(r_1) = f(r_2)$. Thus, f is not injective.

Case 2. The element s_{i_0} has exactly one pre-image $r_{i_0} \in R$. In this case, let us introduce a new function $g : R \setminus \{r_{i_0}\} \to S \setminus \{s_{i_0}\}$ defined by $g(r) = f(r)$ for every $r \in R \setminus \{r_{i_0}\}$. But then $\text{card}\left(R \setminus \{r_{i_0}\}\right) > \text{card}\left(S \setminus \{s_{i_0}\}\right) = k$. Therefore, by our inductive hypothesis, g is not injective. Hence, $g(r_1) = g(r_2)$ for some distinct elements $r_1 \in R$ and

$r_2 \in R$, which means that $f(r_1) = f(r_2)$. Thus, f is not injective.

Case 3. The element s_{i_0} has two or more pre-images. In this case, clearly there are at least two distinct elements, say $r_1 \in R$, and $r_2 \in R$, such that $f(r_1) = f(r_2)$. Thus, f is not injective.

And the entire proof is thereby completed through a combination of the **methods of mathematical induction, and exhaustion.** ♣

Exercise 7: Express the contrapositive of Theorem 6.

Solution
The contrapositive of Theorem 6 may be written as:
Assume R and S are nonempty finite sets. If there is an injection from R into S, the $\text{card}(R) \leq \text{card}(S)$. ♥

Our next theorem is similar to the Pigeonhole principle in that it gives us the condition that the sizes of two finite sets R and S should obey for a surjection to exists from R onto S.

Theorem 8: Assume R and S are nonempty finite sets. If $\text{card}(R) < \text{card}(S)$, then there is no surjection from R onto S.

Proof
We shall again use **mathematical induction** on the size of the codomain set S.

So, let us denote by $P(n)$ the open sentence:

If $\text{card}(R) < \text{card}(S) = n$, then there is no surjection from R onto S.

Initial step. From the above, we see that $P(2)$ is the proposition:

If R contains exactly one item, say $R = \{r\}$, and S exactly two items, then there is no surjection from R into S.

Clearly $P(2)$ is true. For, indeed, let $f : R \to S$ be any function from R into S, then that element of S which is different from $f(r)$ admits no pre-image in R. Thus, f is not surjective.

Induction hypothesis. Assume $P(k)$ is true for some natural number k. That is, assume the proposition

If $\mathrm{card}(R) < \mathrm{card}(S) = k$, then there is no surjection from R onto S

is true for some natural number k.

Inductive step. Now, we must show that the truth of $P(k+1)$ follows from the foregoing induction hypothesis.

Let $f : R \to S$ and $\mathrm{card}(R) < \mathrm{card}(S) = k+1$. Since S has size $k+1$, we list it as

$$S = \{s_1, s_2, s_3, \ldots, s_{k+1}\}$$

Now, let us fix in S an item, say s_{i_0}. We distinguish two cases:

Case 1: $f^{-1}(\{s_{i_0}\}) = \varnothing$. Therefore s_{i_0} has no pre-image in S. Hence, f is not surjective.

Case 2: $f^{-1}(\{s_{i_0}\}) \neq \varnothing$. In this case, let us consider the following function:

$g : R \setminus f^{-1}(\{s_{i_0}\}) \to S \setminus \{s_{i_0}\}$ defined by $g(r) = f(r)$ for every $r \in R \setminus f^{-1}(\{s_{i_0}\})$.

But then, $\mathrm{card}\left(R \setminus f^{-1}(\{s_{i_0}\})\right) < \mathrm{card}\left(S \setminus \{s_{i_0}\}\right) = k$. Therefore, by our inductive assumption, g is not surjective. Hence, there is in $S \setminus \{s_{i_0}\}$ an item, say s, that admits no pre-image in $R \setminus f^{-1}(\{s_{i_0}\})$ by g. Therefore, that same element s cannot admit a pre-image by f. Thus, f is not surjective.

And so the proof is completed through a combination of the **methods of mathematical induction, and exhaustion** . ♣

Exercise 9: Express the contrapositive of Theorem 8.

Solution
The contrapositive of Theorem 8 may be written as:
Assume R and S are nonempty finite sets. If there is a surjection from R onto S, the $\text{card}(R) \geq \text{card}(S)$. ♥

By combining the results of Exercises 7 and 9, we immediately arrive at the following.

Corollary 10: Assume R and S are nonempty finite sets. If there is a bijection $f : R \to S$, then $\text{card}(R) = \text{card}(S)$.

Proof
We shall leave it to the reader to supply the details. ♣

With our next theorem, we see that not only does Corollary 10 hold, but so too does its converse.

Theorem 11: Assume R and S are nonempty finite sets. Then, there is a bijection $f : R \to S$, if and only if $\text{card}(R) = \text{card}(S)$.

Proof
Clearly, we need only establish the converse of Corollary 10. So, suppose $\text{card}(R) = \text{card}(S) = n \in \mathbb{N}$, where n is the common value of these two cardinalities. Since R and S are nonempty finite sets, there exists bijections:

$$f_1 : \{1, 2, 3, \ldots, n\} \to R \quad \text{and} \quad f_2 : \{1, 2, 3, \ldots, n\} \to S.$$

Furthermore, we know that the inverse function of a bijection is also a bijection. Therefore,

$$f_1^{-1} : R \to \{1, 2, 3, \ldots, n\}$$

is a bijection. Finally, since the composition of two bijections is also a bijection, we immediately conclude that $f_2 \circ \left(f_1^{-1}\right) : R \to S$ is bijective. And the proof is complete. ♣

Remarks 12: In other words, for a bijection to exist between two nonempty finite sets it is sufficient and necessary that the two sets have exactly the same number of elements. ■

Corollary 13: Assume A is a nonempty finite set, and assume S is any proper subset of A. Then there is no bijection between A and S.

Proof

This corollary is a direct consequence of Theorems 5 and 11. ♣

Remark 14: As we shall see in the next section, Corollary 13 applies only to finite sets. More precisely, we will show that given a nonfinite [or infinite] set A, one can always establish a bijection between A and one of its proper subsets S. **This seems somehow counter-intuitive**; however, that is a fact we shall soon establish. ■

18.2. INFINITE SETS

In this section we will study infinite sets. Hence, it is a good idea to start with a working definition:

Definition 15 [Infinite sets]: A set S is said to be **infinite**, if it is simply not finite.

For instance, we showed in Example 4.2 that the set of natural numbers $\mathbb{N}$ is not finite. Thus, according to the present definition, $\mathbb{N}$ is an infinite set.

In the preceding section we introduce the notion of *cardinality* for finite sets. And we showed that the cardinality of a nonempty finite set is always greater than that of any of its proper subsets. This result, which holds for finite sets, is however not true in the context of infinite sets. For example, even though the set $\mathbb{E} = \{2, 4, 6, \ldots\}$ of even numbers is clearly a proper subset of the set $\mathbb{N} = \{1, 2, 3, \ldots\}$ of natural numbers, we will see that both sets have the same cardinality [or size] in a sense that is defined next.

Definition 16 [Equinumerous sets]: Let S and R be any two sets. Then S and R are said to have the same cardinality or to be **equinumerous**, and we write $\text{card}(R) = \text{card}(S)$, if there exists a bijection $f : R \to S$.

Remarks 17:

1. In light of Theorem 11, it is clear that the present definition is also consistent with the case where R and S are finite sets. Therefore, it provides an extension of the notion of cardinality to infinite sets.

2. Also, note that equinumerosity is a symmetric relation. For, indeed we know that if there is a bijection $f : R \to S$, then so too is there a bijection $f^{-1} : S \to R$. [In fact, one can easily see that equinumerousity is an equivalent relation. Prove it.] ■

Example 18: Show that

$$\mathbb{E} = \{2, 4, 6, \ldots\} \text{ and } \mathbb{N} = \{1, 2, 3, \ldots\}$$

are equinumerous sets.

Solution

According to Definition 16, to show that $\mathbb{E}$ and $\mathbb{N}$ are equinumerous, it suffices that we construct a bijection between the two sets, which is not difficult to do. Indeed, one readily verifies that the function

$$f : \mathbb{N} \to \mathbb{E} \quad \text{such that} \quad f(n) = 2n$$

is a bijection from $\mathbb{N}$ into $\mathbb{E}$. ♥

Exercise 19: Show that equinumerosity is an equivalent relation.

Solution

Since this is a trivial exercise, we leave it to the reader to supply all the details. ♥

Now, we must be very careful: because the two infinite sets $\mathbb{E}$ and $\mathbb{N}$ are equinumerous does not mean that all other pairs of infinite sets are equinumerous. This is shown by the following exercise.

Exercise 20: Show that the set of natural numbers $\mathbb{N}$ and the interval of real numbers $I =]0, 1[$ are not equinumerous.

Solution

We will conduct a proof **by contradiction** and **construction**. But before we begin the proof proper, we need to recall the following fact learned in high school:

Every real number $x \in I$ admits a unique decimal representation of the form

$$x = 0.d_1\, d_2\, d_3 \ldots, \qquad \text{with } d_k \in \{0, 1, 2, 3, \ldots, 9\},$$

provided it is agreed that if in such expansion all the digits d_n with n greater than some index N are nine but x_N is not a nine, then we shall replace those digits with zeros and replace d_N with $d_N + 1$.

Now, let us assume to the contrary that $\mathbb{N}$ and I are equinumerous. This would mean that there exists a bijection $f : \mathbb{N} \to]0, 1[$. This in turn would mean that all the elements of the interval $I =]0, 1[$ are the images of a sequence. Thus, the totality of the set $I =]0, 1[$ may be listed as

$$x_1$$

$$x_2$$

$$x_3$$

$$\vdots$$

or as follows, [if we take into account the decimal expansion of these numbers]

$$\begin{aligned} x_1 &= 0\,.\,d_{11}\,d_{12}\,d_{13}\,.\,.\,. \\ x_2 &= 0\,.\,d_{21}\,d_{22}\,d_{23}\,.\,.\,. \\ x_3 &= 0\,.\,d_{31}\,d_{32}\,d_{33}\,.\,.\,. \\ &. \\ &. \\ &. \end{aligned} \tag{3}$$

Now, let us introduce [by pure construction] the following number:

$$a = 0\,.\,a_1\,a_2\,a_3\,.\,.\,. \tag{4}$$

where

$$a_k = \begin{cases} d_{kk} + 1\,, & \text{if } 0 \le d_{kk} < 5 \\ d_{kk} - 1\,, & \text{if } 5 \le d_{kk} \le 9 \end{cases}$$

Thus, for every $k \in \mathbb{N}$, the digit a_k is different from both 9 and 0. Therefore, (4) is both an acceptable expansion and an element of the interval I. Furthermore, we have

$$\forall\, k \in \mathbb{N},\ \ a_k \neq d_{kk}\,.$$

Hence (4) must differ from every one of the expansions listed in (3). Therefore, the list (3) does not include (4), which clearly contradicts the assumption that the totality of the set I may be listed as (3). This completes the proof. ♥

Exercise 21: Show that the interval $I =]0,\ 1[$ of real numbers is not a finite set.

Solution

The proof in this exercise is almost identical to that in Exercise 20. Hence, we encourage the reader to supply all the details by himself or herself. ♥

In Exercise 20, we established the fact that, even though $\mathbb{N}$ and $I =]0, 1[$ are both infinite sets, they are not equinumerous. Thus, by Definition 16, these two infinite sets have different cardinalities. This means that there are at least two different **infinite cardinalities** or **transfinite numbers**. In fact, as is shown by our next theorem, the number of transfinite numbers, or infinite cardinalities, is itself infinite. But before presenting that theorem, we will need the following

Definition 22: Let R and S be any two sets.

We shall say that the cardinality of R is less than or equal to the cardinality of S, and we write $\text{card}(R) \leq \text{card}(S)$, if and only if R and **some** subset of S are equinumerous.

And we shall say that the cardinality of R is less than the cardinality of S, and we write $\text{card}(R) < \text{card}(S)$, if and only if these two conditions are satisfied:

1. R and **some** subset of S are equinumerous,
2. but R and S are not equinumerous.

Remark 23: At this point, it may be well to know that there exists an important result named **Schroeder-Berstein Theorem**. This result affirms that if any two sets R and S are such that $\text{card}(R) \leq \text{card}(S)$ and $\text{card}(S) \leq \text{card}(R)$, then the two sets are equinumerous. Clearly, this is a true result in the case where S and R are finite sets. Thus the Schroeder-Berstein result is the extension to infinite sets of a result we already know to be true in the case of finite sets. It is the only major result we mention in this book but do not take time to prove. For, its proof, though not difficult, is lengthy and tedious. ■

We are now in the position to show the following theorem which guarantees the existence of an unending sequence of transfinite numbers.

Theorem 24: Let S be any set. Then $\text{card}(S) < \text{card}(2^S)$.

Proof

We will distinguish two cases:

Case 1. The set S is finite:

In this case, by recalling from Theorem 15 of Chapter 4 the fact that

$$\text{card}\left(2^S\right) = 2^{\text{card}(S)},$$

then all we need to show is that

$$\text{card}(S) < 2^{\text{card}(S)}.$$

But, this latter result was proved by induction on card(S) in Example 10 of Chapter 12.

Case 2. The set S is infinite:

Clearly $\text{card}(S) \le \text{card}\left(2^S\right)$, since there exists an obvious bijection between S and the set of all **singletons**[22] of 2^S. For the remainder of the proof, we will argue by **contradiction**. So, let us assume

$$\text{card}(S) = \text{card}(2^S). \tag{5}$$

Then, according to Definition 16, S and 2^S are equinumerous. Hence, there is a bijection $f : S \to 2^S$. This means

$$\forall A \in 2^S, \ \exists! s \in S, \ f(s) = A. \tag{6}$$

Now, let us consider the set K defined by

$$\text{K} = \{ x \in S : x \notin f(x) \}. \tag{7}$$

Clearly, K is a subset of S. Hence, K must have a unique pre-image y in S :

$$f(y) = \text{K}. \tag{8}$$

Now, we want to know whether the element y belongs to K or $S - \text{K}$. [Recall that it cannot be in both sets.]

22 A **singleton** is any set that contains a single element.

If $y \in K$, then from (7) and (8) we have $y \notin f(y) = K$. In other words, we would have both $y \in K$ and $y \notin K$ which is clearly a contradiction.

On the other hand, if $y \notin K$ then from (8) we have $y \notin f(y)$. Therefore, using (7), we deduce that $y \in K$. Thus, again we have arrived at the contradiction that $y \notin K$ and $y \in K$.

And the proof is therefore complete. ♣

Remark 25:
An important consequence of the preceding theorem is that if S is an infinite set, then

$$\text{card}(S), \quad \text{card}(2^S), \quad \text{card}\left(2^{2^S}\right), \quad \ldots$$

is an entire sequence of transfinite numbers, each one strictly larger than the one before it. ■

Thus, a good question to ask is: which one of all the transfinite numbers is the smallest? The answer to this question will be given by our next theorem. But in the meantime, we want to introduce another:

Definition 26: Let S be a set. Then S is said to be **countably infinite**, if $\mathbb{N}$ and S are equinumerous. On the other hand, we shall say that S is **countable**, if it is either countably infinite or finite. Finally, we shall call S **uncountable**, if it is not countable; that is to say, if it neither countably infinite nor finite.

For example, in light of the results of Exercises 20 and 21, we know that the interval $I =]0, 1[$ is neither finite nor countably infinite. $I =]0, 1[$ is therefore an uncountable set. In contrast, from the results of Example 18 of the present chapter and Example 19 of Chapter 15, we see that the set $\mathbb{E}$ of even numbers and the set $\mathbb{O}$ of odd numbers are both countably infinite.

Remark 27: By convention, the cardinality of countably infinite sets is named **Aleph null** after the first letter $\aleph_0$ in the Hebrew Alphabet. As for the cardinality of the interval $I =]0, 1[$ of real numbers [or of any set which is equinumerous with I], it is called the **cardinality of the continuum** and is customarily denoted by the letter C. ■

With the foregoing definition, we are now ready to establish the following theorem which shows that $\aleph_0$ is the least transfinite numbers.

Theorem 28: Every infinite set has a countably infinite subset.

Proof

Let S be an infinite set. From S we pick an element s_1, and we let $S_1 = S - \{s_1\}$. Clearly, since S is infinite, so too is $S_1 = S - \{s_1\}$. Hence, from $S_1 = S - \{s_1\}$ we can pick a new element s_2 and let S_2 be the infinite set $S_2 = S_1 - \{s_2\} = S - \{s_1, s_2\}$. Continuing this way, we see that after picking the subset $\{s_1, s_2, s_3, \ldots, s_n\}$, the set $S_n = S - \{s_1, s_2, s_3, \ldots, s_n\}$ of remaining elements is still infinite. Hence, we can still choose a new element s_{n+1} from $S_n = S - \{s_1, s_2, s_3, \ldots, s_n\}$. Now, it is not difficult to see that the above process **defines recursively** a sequence $(s_n)_{n=1}^{\infty}$ of elements of S. Also, we note that the terms of this sequence are **by construction** all distinct. Thus, the set $S_0 = \{s_1, s_2, s_3, \ldots\}$, made up of these terms of our sequence, is a countably infinite subset of S. And that completes our proof. ♣

Our next theorem gives another characterization of infinite sets. In fact, it has been adopted by some writers as the definition of an infinite set.

Theorem 29: A set S is infinite, if and only if S is equinumerous with a proper subset of itself

Proof

To prove this theorem, we must establish two implications:

1. S is infinite $\Rightarrow$ S is equinumerous with a proper subset of itself.

2. S is equinumerous with a proper subset of itself $\Rightarrow$ S is infinite.

Here are the proofs of these implications:

1. Assume S is infinite. Then, by Theorem 28, S contains a countably infinite subset S_0:

$$S_0 = \{s_1, s_2, s_3, \ldots\} \subset S.$$

Let $S_1 = S_0 - \{s_1\}$:

$$S_1 = \{s_2, s_3, s_4, \ldots\} \subset S.$$

Clearly, S_0 and S_1 are equinumerous; for indeed, we have the bijection:

$$f : S_0 \to S_1 \quad \text{such that } f(s_n) = s_{n+1}, \quad \forall n \in \mathbb{N}.$$

Similarly, let us introduce another set S^* as $S^* = S - \{s_1\}$. Clearly, S^* is a proper subset of S.

Now, it is also easy to see that the function $I : (S - S_0) \to (S^* - S_1)$, defined by $I(x) = x$ for every $x \in S - S_0$, is the identity function on $S - S_0$; for indeed, $S - S_0 = S^* - S_1$. Hence, I is also bijective.

To finish, we will show that there exists a bijection F on the set S onto its proper subset S^*. Note that to get such a bijection F, it suffices to simply apply the **Gluing Principle** to the functions f and I. Doing so we obtain:

$$F(x) = \begin{cases} f(x) = s_{n+1}, & \textit{if } x = s_n \in S_0, \\ I(x) = x, & \textit{if } x \in S - S_0. \end{cases}$$

Therefore S and its proper subset S^* are equinumerous.

2. As for this implication, it is by contraposition equivalent to:

> 3*. S is finite $\Rightarrow$ S is not equinumerous with any proper subset of itself.

But 3* is precisely Corollary 13 whose truth has already been established. Hence, the proof of the entire theorem is complete. ♣

Finally, by contraposition of Theorem 29, we arrive at the following characterization of finite sets.

Theorem 30: A set S is finite, if and only if S is not equinumerous with any of its proper subsets.

PROBLEMS

1. Show that the intervals of real numbers $I =]0,\ 1[$ and $\mathbb{R}^+ =]0,\ \infty[$ are equinumerous.

2. Show that the intervals of real numbers $I =]0,\ 1[$ and $J =]A,\ B[$ are equinumerous; we will assume $A \in \mathbb{R}$, $B \in \mathbb{R}$ and $A < B$.

3. Show that the intervals of real numbers $I =]A,\ B[$ and $J = [A,\ B[$ are equinumerous; we will assume $A \in \mathbb{R}$, $B \in \mathbb{R}$ and $A < B$.

4. Show that the intervals of real numbers $I =]0,\ 1[$ and $\mathbb{R} =]-\infty,\ \infty[$ are equinumerous.

5. * Let A be any fixed set and let Γ be the function that assigns to each subset D of A its characteristic function χ_D . In other words, Γ is the function defined by:

$$\Gamma : 2^A \to \{0, 1\}^A \quad \text{such that} \quad \Gamma(D) = \chi_D,$$

where $\{0, 1\}^A$ stands for the set of all functions $f : A \to \{0,1\}$. Prove that Γ is a bijection. Deduce from this that $\text{card}(2^A) = \text{card}\left(\{0, 1\}^A\right)$.

"Anything that helps communication is good. Anything that hurts it is bad. I like words more than numbers, and I always did conceptual more than computational."
Paul Halmos.

CHAPTER 19

INDEXED FAMILIES OF SETS

As you may recall, in Chapter 4, we define a set as a collection of **distinguishable** items. Hence, strictly speaking, an aggregate such as $\{a, a, b\}$ is not really a set, since in its listing the same element a appears more than once. The collection $\{a, a, b\}$ is an example of what mathematicians call a **family** or a **multi-set**. In this chapter, we will focus on those families whose elements are themselves sets. We call such families: **families of sets**. Later, we will focus our attention on a particular kind of families of sets called **indexed families** of sets.

19.1. FAMILIES OF SETS

Definition 1 [Family of Sets]: Let U be a fixed set. Then any nonempty collection of subsets of U is called a **family** of sets, or more precisely a **family of subsets of** U.

Example 2: Tell which ones of the following statements are true.

1. Let $U = \mathbb{N}$. Then $F = \{\mathbb{E}, \mathbb{O}\}$ is a family of subsets of U.
2. Let U be any set. Then $F = 2^U$ is a family of subsets of U.

3. Let $U = \{ 1, 2, 3, 4, 5 \}$.

Then $F = \{ \{2, 3\}, \{3, 1\}, \{x : x^2 - 5x + 6 = 0\}, \{2, 3, 5\} \}$ is a family of subsets of U.

4. Let $U = \{ 2, 3, 4, 5 \}$.

Then $F = \{ \{2, 3\}, \{0, 3, 1\}, \{x : x^2 - 5x + 6 = 0\}, \{2, 3, 5\} \}$ is a family of subsets of U.

5. Let $U = \mathbb{Q}$. Then $F = \{ \{ x \in \mathbb{Q}, x < i\} : i \in \mathbb{R} \}$ is a family of subsets of U.

Solution

1. True.

2. True.

3. True. Also you may note that the family F includes two indistinguishable elements, namely:

$$\{2, 3\} \text{ and } \{x : x^2 - 5x + 6 = 0\}.$$

4. False. Even though F is a family of sets, it is not a family of subsets of U, for indeed the element $\{0, 3, 1\}$ is not a subset of the set U.

5. True. ♥

19.2. INDEXED FAMILIES OF SETS

To introduce the idea of an **indexed family of sets**, let us consider the following simple game. In this game, the set $I = \{g, r, e, a, t\}$ is given. And the game consists in associating with each element $i \in I$ the subset S_i of all consonants in I that are different from the element i itself. Thus, we have:

$$S_g = \{r, t\}, \quad S_r = \{g, t\}, \quad S_e = \{g, r, t\}, \quad S_a = \{g, r, t\}$$

and $S_t = \{g, r\}$.

Note that S_e and S_a are exactly the same set. Hence, what we have arrived at in this exercise is the collection

$$F = \{ S_g, \quad S_r, \quad S_e, \quad S_a, \quad S_t \}$$

of sets each of which is associated with an element of I. We call F a family of sets indexed by I.

Definition 3 : Let I and U be any given sets. If for every $i \in I$ there corresponds a subset S_i of U, then the collection

$$F = \{ S_i : \; i \in I \},$$

whose elements S_i may or may not be all distinct, is called an **indexed family** of subsets of U. The set I is the **indexing set**, and the variable $i \in I$ the **index**.

Notice that in cases where the indexing set I is the set $\mathbb{N}$ of natural numbers, the indexed family of sets F is simply a sequence of sets that may be written more compactly as $F = (S_n)_{n=1}^{\infty}$.

Example 4: Given the following family of sets

$$F = \{ \{ r \in \mathbb{Q}, \; r < x \} : x \in \mathbb{R} \},$$

identify

1. U, 2. I, 3. i, 4. And S_i

Solution

1. $U = \mathbb{Q}$, 2. $I = \mathbb{R}$, 3. $i = x$,

4. and $S_i = S_x = \{ r \in \mathbb{Q}, \; r < x \}$. ♥

Example 5: Given the following family of sets

$$F = \left\{ \left[0, \tfrac{1}{n} \right] : n \in \mathbb{N} \right\},$$

identify

1. U, 2. I, 3. i, 4. and S_i

Solution

1. $U = \mathbb{R}$, 2. $I = \mathbb{N}$, 3. $i = n$,

4. and $S_i = S_n = \left[0, \frac{1}{n} \right]$.

Also, since $I = \mathbb{N}$, note that $F = \left\{ \left[0, \frac{1}{n} \right] : n \in \mathbb{N} \right\}$ may be rewritten as $F = \left(\left[0, \frac{1}{n} \right] \right)_{n=1}^{\infty}$. ♥

Remark 6: In the remainder of this chapter we will consider only indexed families. The reason for this is simple: Indeed, if F is any family of sets, with an arbitrarily large cardinality, then it is always possible to introduce as its indexed set any set I having the same cardinality as F, and to assume that the elements of F have been indexed by I. In other words, it is a convenient thing to view every family of sets as an indexed family [of sets]. ■

19.3. OPERATIONS ON INDEXED FAMILIES

You may recall that in Chapter 5 we defined the operations of intersection, union, complementation, and Cartesian product for pairs of sets. In the present section we will extend these operations to indexed families of sets.

Definition 7 [Operations on an Indexed Family of Sets]: Assume $F = \{ S_i : i \in I \}$ is an indexed family of subsets of a universal set U. Then by the **union, intersection** and **Cartesian product** of the elements of F, we mean, respectively:

$$\bigcup_{i \in I} S_i = \{ x : \exists i \in I, \ x \in S_i \}, \tag{1}$$

$$\bigcap_{i \in I} S_i = \{ x : \forall i \in I, \ x \in S_i \}, \tag{2}$$

and

$$\times_{i \in I} S_i = \{ (x : I \to \cup_{i \in I} S_i) \text{ such that } \forall i \in I, \ x(i) \in S_i \} \tag{3}$$

In other words, for an object to be in the union of a family of sets, it is **necessary and sufficient** that the object belongs to at least one of the sets constituting the family. Whereas, for an object to be in the intersection of a family of sets, that object must belong simultaneously to all of the sets forming the family.

Remark 8: As for $\times_{i\in I} S_i$, the Cartesian product of a family of sets, it is a more subtle notion. Intuitively, one may think of $\times_{i\in I} S_i$ as the collection of all "generalized" lists $(x_i)_{i\in I}$ such that $x_i \in S_i$, $\forall i \in I$; but rigorously speaking, we define $\times_{i\in I} S_i$ as the set of all functions

$$x : I \to \bigcup_{i\in I} S_i \quad \textit{such that} \quad \forall i \in I,\ x(i) \in S_i.$$

Any one of these functions is called a **choice function**. That such a function exists for any nonempty family $\{ S_i :\ i \in I\}$ of nonempty sets S_i is an axiom of set theory known as **the axiom of choice**. ■

Remark 9: Perhaps you have noticed that we did not exclude from Definition 7 empty families $F = \{ S_i :\ i \in I\}$. This was done deliberately. For, in the case where $F = \{ S_i :\ i \in I\}$ is an empty family, that is if $F = \{ S_i :\ i \in I\}$ contains no set S_i at all, then it is customary to adopt the following **convention**:

$$\bigcup_{i\in I} S_i = \varnothing \quad \textit{and} \quad \bigcap_{i\in I} S_i = U$$

where U is our universal set. Though this may seem like a **completely arbitrary convention**, we can make some sense of it, which in turn will help us commit the cconvention to memory more easily. To make sense of it, we reason as follows. *If* $F = \{ S_i :\ i \in I\}$ *is an empty family, that is to say a family with absolutely no set* S_i*, then* $I = \varnothing$*. Therefore no object* x *satisfies the defining property,* $\exists i \in I,\ x \in S_i$*, of the union of the elements of* F*. Hence it is reasonable to allow* $\bigcup_{i\in I} S_i = \varnothing$*. Similarly, when* $F = \{ S_i :\ i \in I\}$ *is an empty family, the defining property,* $\forall i \in I,\ x \in S_i$*, for the intersection of the elements of* F*, which may be rewritten as*

$\forall i,\ i \in I \rightarrow x \in S_i$, *is a vacuously true sentential implication. Therefore, we can let* $\cap_{i \in I} S_i = U$. ■

We can now procede to establish our first theorem. This theorem is called the **DeMorgan's laws of set theory**, since it is the analog of the DeMorgan's laws of propositional logic. Thus, as its name suggests, the theorem comes in two parts:

1. one that states that the intersection of the complements of a family of sets is the complement of their union.
2. and another that says that the union of the complements of a family of sets is the complement of their intersections.

Theorem 10 [DeMorgan's Laws]: Let U the universal set, and A a fixed subset of U. If $\{\ S_i:\ i \in I\}$ is an indexed family of subsets of U, then:

$$\cap_{i \in I} (A - S_i) = A - (\cup_{i \in I} S_i), \qquad (4)$$

and

$$\cup_{i \in I} (A - S_i) = A - (\cap_{i \in I} S_i). \qquad (5)$$

Proof
We shall give only the proof of (4), since that of (5) is obtained pretty similarly. So too we encourage the reader to write out the proof of (5) with all its details.

We know that (4) is equivalent to:

$$x \in \left(\cap_{i \in I}(A - S_i)\right) \iff x \in (\ A - (\cup_{i \in I} S_i)\ ) \qquad (6)$$

Therefore, to establish (4), it suffices that we begin with the right-hand sight of (6), and proceed by successive equivalences, checking both sides of each equivalence, till we reach the left-hand side:

$$x \in \left(\cap_{i \in I}(A - S_i)\right) \iff \forall i \in I,\ \ x \in (A - S_i)$$
$$\iff \forall i \in I,\ \left(x \in A \ \ and \ \ x \notin S_i\right)$$

$$\Leftrightarrow \quad x \in A \quad \textit{and} \quad (\forall i \in I,\ x \notin S_i)$$

$$\Leftrightarrow \quad x \in A \quad \textit{and} \quad x \notin (\cup_{i\in I} S_i)$$

$$\Leftrightarrow \quad x \in (A - (\cup_{i\in I} S_i)) . \clubsuit$$

To close this section we give one more

Theorem 11: Let $\mathbb{A}$ and $\mathbb{B}$ be two nonempty sets, and $f : \mathbb{A} \to \mathbb{B}$ a function. Also, let $(A_i)_{i\in I}$ and $(B_i)_{i\in I}$ be families of subsets of $\mathbb{A}$ and $\mathbb{B}$, respectively. Then,

1. $f(\varnothing) = \varnothing$.
2. $f^{-1}(\varnothing) = \varnothing$.
3. $f^{-1}(\cup_{i\in I} B_i) = \cup_{i\in I} f^{-1}(B_i)$.
4. $f^{-1}(\cap_{i\in I} B_i) = \cap_{i\in I} f^{-1}(B_i)$.
5. $f(\cup_{i\in I} A_i) = \cup_{i\in I} f(A_i)$.
6. $f^{-1}(\mathbb{B} - B_i) = \mathbb{A} - f^{-1}(B_i)$.
7. $f(\cap_{i\in I} A_i) \subset (\cap_{i\in I} f(A_i))$.
8. $A_i \subset (f^{-1} \circ f(A_i))$.
9. $f \circ f^{-1}(B_i) \subset B_i$
10. $f(\mathbb{A} - A_i) \subset \mathbb{B}$.

Proof

1. We will prove this proposition by **contradiction**. Suppose $f(\varnothing)$ is not the empty set.

$$f(\varnothing) \neq \varnothing \quad \Leftrightarrow \quad \exists b \in \mathbb{B},\ b \in f(\varnothing)$$
$$\Leftrightarrow \quad \exists a \in \varnothing,\ b = f(a),$$

which contradicts the fact that there is no element in the empty set $\varnothing$.

2. We will prove this proposition too by **contradiction**. Suppose $f^{-1}(\varnothing)$ is not the empty set

$$f^{-1}(\varnothing) \neq \varnothing \quad \Leftrightarrow \quad \exists a \in \mathbb{A},\ a \in f^{-1}(\varnothing)$$
$$\Leftrightarrow \quad \exists a \in \mathbb{A},\ f(a) \in \varnothing,$$

which contradicts the fact that $\varnothing$ is empty.

3. To prove this proposition we will use the **method of direct proof** with **equivalences:**

$$\begin{aligned} x \in f^{-1}(\cup_{i\in I} B_i) &\Leftrightarrow f(x) \in \cup_{i\in I} B_i \\ &\Leftrightarrow \exists i \in I,\ f(x) \in B_i \\ &\Leftrightarrow \exists i \in I,\ x \in f^{-1}(B_i) \\ &\Leftrightarrow x \in \left(\cup_{i\in I} f^{-1}(B_i)\right) \end{aligned}$$

4. To prove this proposition we will use the **method of direct proof** with **equivalences:**

$$\begin{aligned} x \in f^{-1}(\cap_{i\in I} B_i) &\Leftrightarrow f(x) \in \cap_{i\in I} B_i \\ &\Leftrightarrow \forall i \in I,\ f(x) \in B_i \\ &\Leftrightarrow \forall i \in I,\ x \in f^{-1}(B_i) \\ &\Leftrightarrow x \in \left(\cap_{i\in I} f^{-1}(B_i)\right) \end{aligned}$$

5. To prove this proposition we will use the **method of direct proof** with **equivalences:**

$$\begin{aligned} y \in f(\cup_{i\in I} A_i) &\Leftrightarrow \exists x \in \cup_{i\in I} A_i,\ y = f(x) \\ &\Leftrightarrow \exists i \in I,\ \exists x \in A_i,\ y = f(x) \\ &\Leftrightarrow \exists i \in I,\ y \in f(A_i) \\ &\Leftrightarrow y \in \left(\cup_{i\in I} f(A_i)\right) \end{aligned}$$

6. To prove this proposition we will use the **method of direct proof** with **equivalences:**

$$\begin{aligned} x \in f^{-1}(\mathbb{B} - B_i) &\Leftrightarrow f(x) \in \mathbb{B} - B_i \\ &\Leftrightarrow f(x) \in \mathbb{B},\ \text{and}\ f(x) \notin B_i \\ &\Leftrightarrow x \in f^{-1}(\mathbb{B}),\ \text{and}\ x \notin f^{-1}(B_i) \\ &\Leftrightarrow x \in \mathbb{A},\ \text{and}\ x \notin f^{-1}(B_i),\ \text{Since}\ f^{-1}(\mathbb{B}) = \mathbb{A} \\ &\Leftrightarrow x \in \left(\mathbb{A} - f^{-1}(B_i)\right). \end{aligned}$$

7. To prove this proposition we will use the **method of direct proof** with **implications:**

$$\begin{aligned} y \in f(\cap_{i\in I} A_i) &\Rightarrow \exists x \in \cap_{i\in I} A_i,\ y = f(x) \\ &\Rightarrow \exists x \in \mathbb{A},\ \forall i \in I,\ x \in A_i\ \text{and}\ y = f(x) \end{aligned}$$

$$\Rightarrow \exists x \in \mathbb{A}, \left(\forall i \in I, \left(x \in A_i \quad and \quad y = f(x)\right)\right)$$

$$\Rightarrow \forall i \in I, \left(\exists x \in \mathbb{A}, \left(x \in A_i \quad and \quad y = f(x)\right)\right),$$

by Theorem 10.2 of Chapter 8

$$\Rightarrow \forall i \in I, \left(\exists x \in A_i \quad and \quad y = f(x)\right)$$

$$\Rightarrow \forall i \in I, \quad y \in f(A_i)$$

$$\Rightarrow y \in \left(\cap_{i \in I} f(A_i)\right)$$

8. To prove this proposition we will use the **method of direct proof** with **implications**:

$$x \in A_i \Rightarrow f(x) \in f(A_i),$$

$$\Rightarrow x \in f^{-1} o f(A_i)$$

9. To prove this proposition we will use the **method of direct proof** with **implications**:

$$y \in f o f^{-1}(B_i) \Rightarrow \exists x \in f^{-1}(B_i), \quad y = f(x)$$

$$\Rightarrow \exists x \in \mathbb{A}, \quad f(x) \in B_i, \quad y = f(x)$$

$$\Rightarrow y \in B_i.$$

10. This part is obvious; but we will still give a formal proof. To prove this proposition we will use the *method of direct proof* with implications:

$$y \in f(\mathbb{A} - A_i) \Rightarrow \exists x \in (\mathbb{A} - A_i), \quad y = f(x)$$

$$\Rightarrow \exists x \in \mathbb{A}, \quad x \notin A_i, \quad y = f(x)$$

$$\Rightarrow \exists x \in \mathbb{A}, \quad y = f(x)$$

$$\Rightarrow y \in \mathbb{B}.$$

And this completes the proofs of all ten propositions. ♣

19.4. FAMILY OF SETS AND CARDINALITY

In Chapter 18 we started our study of cardinalities. We saw that the set of even numbers $\mathbb{E}$ is equinumerous with $\mathbb{N}$, and therefore has the cardinality of Aleph null. We also proved that $\mathbb{N}$ and the set $I =]0, 1[$

are not equinumerous. In the present section we will continue to study the cardinalities of infinite sets. In each of the theorems that follow, we will consider a family of sets, and we will determine the cardinality for the union of the sets in the family.

19.4. 1. COUNTABLY INFINITE SETS

Theorem 12 [The First Hilbert Hotel Theorem[23]] Let S be a countably infinite set. Then the union $S \cup \{a\}$ is also countably infinite.

In other words, the union of a countably infinite set with any singleton is also countably infinite.

Proof

There are two cases:

Case 1. $a \in S$. In this case, since $S \cup \{a\} = S$ and S is by assumption countably infinite, then so too is $S \cup \{a\}$.

Case 2. $a \notin S$. In this case, let us list the set S in the form $S = \{ s_1, s_2, s_3, \ldots \}$ and introduce the function $f : S \to S \cup \{a\}$ defined by:

$$f(s_k) = \begin{cases} a, & \text{if } k = 1 \\ s_{k-1}, & \text{if } k \geq 2. \end{cases}$$

Clearly, f is a bijection between S and $S \cup \{a\}$. For indeed, every element of $S \cup \{a\}$ has a unique pre-image in S. Thus, S and $S \cup \{a\}$ are equinumerous. Therefore, $S \cup \{a\}$ too is countably infinite, and the proof is complete. ♣

The next theorem is a simple generalization of the preceding one.

23 The reader is encouraged to research with the internet or other means the origin of the name *The Hilbert Hotel Theorerm* given to this result.

Theorem 13: Let S be a countably infinite set and $\{a_1, a_2, a_3, \ldots, a_n\}$ a finite set. Then the union $S \cup \{a_1, a_2, a_3, \ldots, a_n\}$ is also countably infinite.

In other words, the union of a countably infinite set with any finite set is also countably infinite.

Proof

We will prove this theorem by **induction** on n. So let $P(n)$ be the open sentence:

> *The union* $S \cup \{a_1, a_2, a_3, \ldots, a_n\}$, *where* S *is countably infinite, is also countably infinite.*

Initial step. From the above we see that $P(1)$ is the proposition:

> *The union* $S \cup \{a_1\}$, *where* S *is countably infinite, is also countably infinite.*

Thus $P(1)$ is true, for $P(1)$ is simply a re-statement of Theorem 12.

Induction hypothesis. Assume $P(k)$ is true for some k. That is, assume

> *The union* $S \cup \{a_1, a_2, a_3, \ldots, a_k\}$, *where* S *is countably infinite, is also countably infinite.*

is true for some k.

Inductive step. Now, we must deduce from the inductive hypothesis that $P(k+1)$ too is true.

But first, let us observe that

$$\begin{aligned} S \cup \{a_1, a_2, \ldots, a_{n+1}\} &= \left(S \cup \{a_1, a_2, \ldots, a_n\}\right) \cup \{a_{n+1}\} \\ &= S^* \cup \{a_{n+1}\} \qquad (7) \end{aligned}$$

where by the induction hypothesis

$$S^* = S \cup \{a_1, a_2, a_3, \ldots, a_n\}.$$

is countably infinite. Therefore, using Theorem 12, we immediately see that $S^* \cup \{ a_{k+1} \}$ is also countably infinite. And so the proof is complete. ♣

Next we would like to show that the union of two countably infinite sets is also a countably infinite set. But to do that we will need the following lemma.

Lemma 14: Every subset of a countably infinite set is either finite or countably infinite.

In other words, no subset of a countably infinite set shall possess a cardinality larger than **Aleph null**.

Proof

Let S be a countably infinite set. Then S may be written in the form:

$$S = \{ a_1 , a_2 , a_3 , \ldots \} . \tag{8}$$

If R is a subset of S, then there are two possibilities or cases:

Case 1. R is finite: In this case there is nothing to prove, and we are done.

Case 2. R is infinite: Then let a_{n_1} be the first element in the sequence (8) such that $a_{n_1} \in R$; let a_{n_2} be the second element in the sequence (8) such that $a_{n_2} \in R$; and in general, let a_{n_k} be the k-th element in the sequence (8) such that $a_{n_k} \in R$. Clearly, we have $a_{n_i} \neq a_{n_j}$ for $i < j$. Therefore, the function:

$$f : \mathbb{N} \to R \qquad \text{defined by} \qquad f(k) = a_{n_k}$$

is injective. Now let us show that f is also surjective. To do that, let us first observe from Problem 4 of Chapter 16 that $n_k \geq k$ for all $k \in \mathbb{N}$; hence the set $\{ a_{n_1} , a_{n_2} , a_{n_3} , \ldots , a_{n_k} \}$ contains all the elements of R that appear in listing (8) as far as a_k. Now, let b be an arbitrary element of R. Since $R \subset S$, we have $b \in S$. Thus, $b = a_p$ for some $p \in \mathbb{N}$. Therefore, with the observation just made, we can immediately conclude that $b \in \{ a_{n_1}, a_{n_2}, a_{n_3}, \ldots, a_{n_p} \}$. Therefore, $b = a_{n_q}$ for some $q \in \mathbb{N}$. That is to say,

$$f(q) = b \quad \text{for some} \quad q \in \mathbb{N}.$$

Hence, f is also surjective, which completes the proof . ♣

Example 15: Show that the Cartesian product $\mathbb{N} \times \mathbb{N}$ is countably infinite.

Solution

Remember that we establish in Exercise 5 of Chapter 15 that there is an injection between $\mathbb{N} \times \mathbb{N}$ and $\mathbb{N}$. Hence, $\mathbb{N} \times \mathbb{N}$ is equinumerous with some subset of $\mathbb{N}$, say S. But, by Theorem 14, S is either finite or countably infinite. Therefore, $\mathbb{N} \times \mathbb{N}$ too is either finite or countably infinite. However, we know clearly that $\mathbb{N} \times \mathbb{N}$ is not finite; thus it is countably infinite. And the proof is complete. ♥

We are now in the position to prove the following

Theorem 16 [The Second Hilbert Hotel Theorem]: Let R and S be two countably infinite sets. Then their union $R \cup S$ is also countably infinite.

Proof

Let R and S be two countably infinite sets. Then, we may list then as:

$$R = \{ a_1, a_2, a_3, \ldots \} \tag{9}$$

and

$$S = \{ b_1, b_2, b_3, \ldots \}. \tag{10}$$

Next, we may distinguish two cases:

Case 1. The intersection of R and S is empty:

Let us split the set S into two new sets $\mathbb{I}_E$ and $\mathbb{I}_O$ defined by:

$$\mathbb{I}_E = \{ b_2,\ b_4,\ b_6,\ .\ .\ .\ ,\ b_{2n},\ .\ .\ .\}$$

and

$$\mathbb{I}_O = \{ b_1,\ b_3,\ b_5,\ .\ .\ .\ ,\ b_{2n-1},\ .\ .\ .\}$$

such that

$$S = \mathbb{I}_E \cup \mathbb{I}_O .$$

Now, let us introduce the functions

$$f_1 : R \to \mathbb{I}_O \quad \text{defined by} \quad f(a_n) = b_{2n-1} \ \text{ for every } n \in \mathbb{N},$$

and

$$f_2 : S \to \mathbb{I}_E \quad \text{defined by} \quad f(b_n) = b_{2n} \ \text{ for every } n \in \mathbb{N}.$$

Clearly f_1 and f_2 are bijections. Hence, by the **Gluing Principle** [see Problem 4 of Chapter 15], we see that the function

$$f : R \cup S \to S \quad \text{defined by} \quad f(x) = \begin{cases} f_1(x), & \text{if } x \in R \\ f_2(x), & \text{if } x \in S \end{cases}$$

is also a bijection. Thus, $R \cup S$ and S share the same cardinality; $\aleph_0$. Therefore, $R \cup S$ is countably infinite.

Case 2. The intersection of R and S is not empty:

But, we know that $R \cup S = R \cup (S - R)$ with the intersection of R and $S - R$ being empty. So, there are two subcases under this case:

- If $S - R$ is finite, then by applying Theorem 13 to R and $S - R$ we immediately see that $R \cup (S - R)$ too is countably infinite. Therefore, $R \cup S$ is countably infinite.

- On the other hand, If $S - R$ is infinite, then by Lemma 14 it is a countably infinite subset of S. Therefore, applying the result of *Case 1* of this proof to R and $S - R$, we again see that

$R \cup (S - R)$ is countably infinite. Therefore, $R \cup S$ is countably infinite.

And the entire proof is complete. ♣

The following corollary is an easy generalization of the preceding theorem.

Corollary 17: Let $\{S_1, S_2, S_3, \ldots, S_n\}$ be a finite family of countably infinite sets. Then the union $\bigcup_{i=1}^{n} S_i = S_1 \cup S_2 \cup S_3 \cup S_n$ is also countably infinite.

Proof

This proof is easily carried out by **mathematical induction** on the size n of family of sets. So, let $P(n)$ be the open sentence.

The union $\bigcup_{i=1}^{n} S_i$ of n countably infinite sets is also countably infinite.

Initial step. From the above we see that $P(1)$ is the proposition

S_1 *is countably infinite.*

Thus, $P(1)$ is true.

Induction hypothesis. Assume $P(k)$ is true for some $k \in \mathbb{N}$. That is, assume

The union $\bigcup_{i=1}^{k} S_i$ of k countably infinite sets is also countably infinite

is true for some $k \in \mathbb{N}$.

Inductive step. Now, we must deduce from the **induction hypothesis** that that $P(k+1)$ too holds. But, first, let us notice the trivial relation

$$\bigcup_{i=1}^{k+1} S_i = \left(\bigcup_{i=1}^{k} S_i \right) \cup S_{k+1}, \tag{11}$$

where

$\bigcup_{i=1}^{k} S_i$ is countably infinite by the induction hypothesis .

Therefore, applying Theorem 16 to $\bigcup_{i=1}^{k} S_i$ and S_{k+1} , we deduce from (11) that $\bigcup_{i=1}^{k+1} S_i$ too is countably infinite. And the entire proof is complete. ♣

The following Theorem is very important in itself.

Theorem 18: Let R and S be two countably infinite sets. Then their Cartesian product $R \times S$ is also countably infinite.

Proof

Given that $\mathbb{N} \times \mathbb{N}$ is countably infinite [see Example 15], it suffices that we prove the following:

$$R \times S \text{ is equinumerous with } \mathbb{N} \times \mathbb{N}.$$

Let us first note that since R and S are both countably infinite, there are bijections $f : \mathbb{N} \to R$ and $g : \mathbb{N} \to S$.

Now, let us introduce the new function $h : \mathbb{N} \times \mathbb{N} \to R \times S$ defined by:

$$h(n, m) = (f(n), g(m)).$$

We will show that h is a bijection, thereby proving that $R \times S$ and $\mathbb{N} \times \mathbb{N}$ are equinumerous sets. So, let (a, b) and (c, d) be any two ordered pairs of $\mathbb{N} \times \mathbb{N}$:

$$\begin{aligned} h(a, b) = h(c, d) &\Leftrightarrow (f(a), g(b)) = (f(c), g(d)) \\ &\Leftrightarrow f(a) = f(c) \text{ and } g(b) = g(d) \\ &\Leftrightarrow a = c \text{ and } b = d, \\ &\quad \text{since } f \text{ and } g \text{ are both injective} \\ &\Leftrightarrow (a, b) = (c, d) \end{aligned}$$

Thus, $h : \mathbb{N} \times \mathbb{N} \to R \times S$ is an injection. To show that h is a surjection, let (N, M) be an arbitrary element of $R \times S$. Then clearly, $N \in R$ and $M \in S$. Hence, since $f : \mathbb{N} \to R$ and $g : \mathbb{N} \to S$ are surjections, we have

$f(n) = N$ *and* $g(m) = M$, for some ordered pair (n, m) in $\mathbb{N} \times \mathbb{N}$

Thus, $h(n, m) = (f(n), g(m)) = (N, M)$; which shows that h is indeed surjective. And this completes the proof. ♣

Corollary 19: Let $\{ S_1, S_2, S_3, \ldots \}$ be a countably infinite family of countably infinite sets. Then the union $\bigcup_{i=1}^{\infty} S_i$ is also countably infinite.

Proof

There are two cases that we can distinguish:

Case 1. The sets S_i are mutually disjoint. By this, we mean any two sets of the family are disjoint: In this case, $\bigcup_{i=1}^{\infty} S_i$ is of maximal cardinality. Hence, by simply indexing the elements of each S_i as follows

$$S_1 = \{s_{11}, s_{12}, s_{13}, \ldots\}$$
$$S_2 = \{s_{21}, s_{22}, s_{23}, \ldots\}$$
$$S_3 = \{s_{31}, s_{32}, s_{33}, \ldots\}$$
$$\vdots$$

it is easy to see that $\bigcup_{i=1}^{\infty} S_i$ and $\mathbb{N} \times \mathbb{N}$ are equinumerous. Therefore $\bigcup_{i=1}^{\infty} S_i$ is countably infinite.

Case 2. The sets S_i are not mutually disjoint: In this case, $\bigcup_{i=1}^{\infty} S_i$ is equinumerous with a subset of $\mathbb{N} \times \mathbb{N}$ that is not finite; therefore $\bigcup_{i=1}^{\infty} S_i$ it is countably infinite. And the entire proof is thereby completed. ♣

Remark 20: We learned in Chapter 18 that $\aleph_0$ is the smallest transfinite number and that:

$$\aleph_0 = \text{card}(\mathbb{N}) < \text{card}(2^{\mathbb{N}}).$$

So, a very natural question that many mathematicians, including Greg Cantor, the founder of set theory, have been asking was: *Is there a cardinal number x such that*

$$\aleph_0 = \text{card}(\mathbb{N}) < x < \text{card}(2^{\mathbb{N}}).$$

Cantor himself conjectured that **no** such cardinal number exists. In other words, he assumed that after $\aleph_0$, the largest transfinite number is $\text{card}(2^{\mathbb{N}})$ which he denoted by $\aleph_1$. This assumption has ever since been famously known in mathematical circles as the **continuum hypothesis**. ■

PROBLEMS

Problems 1 to 10 deal with countable sets. Hence, we remind the reader that a countable set is a set that is either finite or countably infinite. [To solve these problems, the reader may use the **method of exhaution**]

1. Prove that if S is a countable set, then $S \cup \{a\}$ too is countable.

2. Prove that if S is a countable set, then $S \cap \{a\}$ too is countable.

3. Prove that if S is a countable set, then $S \cap \{a_1, a_2, a_3, \ldots, a_n\}$ too is countable.

4. Prove that if S is a countable set, then $S \cup \{a_1, a_2, a_3, \ldots, a_n\}$ too is countable. [You may use mathematical induction.]

5. Prove that if S and R are countable sets then $S \cup R$ too is countable

6. Prove that if S and R are countable sets then $S \cap R$ too is countable

7. Prove that if $(S_i)_{i=1}^{n}$ is a finite collection of countable sets, then

$\bigcup_{i=1}^{n} S_i$ too is countable. [You may use mathematical induction.]

8. Prove that if $(S_i)_{i=1}^{n}$ is a finite collection of countable sets, then $\bigcap_{i=1}^{n} S_i$ too is countable.

9. Prove that if $(S_i)_{i=1}^{\infty}$ is a countably infinite collection of countable sets, then $\bigcup_{i=1}^{\infty} S_i$ too is countable.

10. Prove that if $(S_i)_{i=1}^{\infty}$ is a countably infinite collection of countable sets, then $\bigcap_{i=1}^{\infty} S_i$ too is countable.

The goal of the remaining problems is to establish the fact that $C = \text{card}(2^{\mathbb{N}})$. In other words, step by step, we will lead you to prove that the power set of any countably infinite set, such as $\mathbb{N}$, has the cardinality of the continuum.

11. Prove that $2^{\mathbb{Q}}$ is equinumerous with $2^{\mathbb{N}}$. [You may use the fact that, $\mathbb{Q}$ being equinumerous with $\mathbb{N}$, every subset of $\mathbb{Q}$ can be paired off with one and only one subset of $\mathbb{N}$, and vice versa.]

12. Consider the function $f : \mathbb{R} \to 2^{\mathbb{Q}}$ defined by

$$f(x) = \{\, r \in \mathbb{Q} : \ r < x \,\}, \qquad \textit{for all } x \in \mathbb{R} .$$

12.1. Prove that f is injective.

12.2. Deduce from 12.1 that $C \leq \text{card}(2^{\mathbb{Q}})$, where $C = \text{card}(\mathbb{R})$.

12.3. Deduce from 12.2 and Problem 11 that $C \leq \text{card}(2^{\mathbb{N}})$.

13. Consider the function $\Sigma : \{0, 1\}^{\mathbb{N}} \to \mathbb{R}$ defined by

$$\Sigma(f) = 0.f(1)f(2)f(3)\cdots \qquad \textit{for all } f \in \{0, 1\}^{\mathbb{N}} .$$

Note that $\Sigma(f)$ is a decimal number consisting of 0's and 1's. Also, note that for any two functions f and $g \in \{0, 1\}^{\mathbb{N}}$, if $f \neq g$ then $\Sigma(f) \neq \Sigma(g)$.

13.1. Is $\Sigma : \{0, 1\}^{\mathbb{N}} \to \mathbb{R}$ injective?

13.2. Deduce from your answer to 13.1 that $\text{card}\left(\{0, 1\}^{\mathbb{N}}\right) \leq C$.

13.3. Deduce from your answers to 13.2 and to Problem 5 of Chapter 18 that $\text{card}(2^{\mathbb{N}}) \leq C$.

13.4. Deduce from your answers to 13.3 and to Problem 12.3 that $C = \text{card}(2^{\mathbb{N}})$. [What important remark made in Chapter 18 allows you to reach this conclusion in a definitive way?]

14. Let $\mathbb{A}$ and $\mathbb{B}$ be two nonempty sets, and $f : \mathbb{A} \to \mathbb{B}$ a bijection. Also, let A and B be subsets of $\mathbb{A}$ and $\mathbb{B}$, respectively. Then prove that

14.1. $A = \left(f^{-1} \circ f(A)\right)$.

14.2. $\left(f \circ f^{-1}(B)\right) = B$.

BIBLIOGRAPY

Creative Psychology of Science and Mathematics

Poincare H., *Science and Hypothesis.* Dover (1952).

General Survey of Mathematics

King, J., *Mathematics in Ten Lessons: The Grand Tour.* Prometheus Books (2009).

Gullberg, J., *Mathematics from the Birth of Numbers.* Norton (1997).

Modern and Linear Algebras

Warner, S., *Modern Algebra.* Dover (1990).

Landin, J., *An Introduction to Algebraic Structures.* Dover (1990).

Number Theory

Ash, A. and Gross, R., Fearless *Symmetry: Exposing the Hidden Patterns of Numbers.* Princeton University Press (2006).

Landin, J., *An Introduction to Algebraic Structures.* Dover (1990).

The Calculus

Anderson, K. W. and Hall, D. W., *Sets, Sequences and Mappings: The Basic Concepts of Analysis.* Dover (2009).

Burkill, J. C., *A First Course in mathematical Analysis.* Cambridge Mathematical Library (2000).

Burkill, J. C. and Burkill H., *A Second Course in mathematical Analysis.* Cambridge Mathematical Library (2002).

The Logic of Mathematics

Bradis, V. M., Minkovskii, V. L., and Kharcheva, A. K., *Lapses in Mathematical Reasoning.* Dover (1999).

Margaris A., *First order Mathematical Logic,* Dover (1990).

INDEX OF MATHEMATICAL TERMS

Made in the USA
Lexington, KY
18 December 2012